Wissenschaft und Philosophie – Science and Philosophy – Sciences et Philosophie

Series Editors

Jürgen Jost, Geometrische Methoden und Komplexe Systeme, Max-Planck-Institut für Mathematik in den Naturwissenschaften, Leipzig, Germany

Martin Carrier, Institute for Interdisciplinary Studies of Science (I2SoS), Universität Bielefeld, Bielefeld, Germany

Die Reihe will den Dialog und die konstruktive Auseinandersetzung zwischen Wissenschaft und Philosophie fördern. Diesem Ziel nähert sie sich von beiden Seiten: Zum einen sollen die Implikationen wissenschaftlicher Ergebnisse und Theorien für das philosophische Denken erkundet werden. Zum anderen soll eine philosophische Durchdringung und Analyse wissenschaftlicher Problemlagen, Konzepte und Theorien erreicht werden. Dabei werden verschiedenartige philosophische Ansätze verfolgt. Die Naturwissenschaften und die Mathematik werden besonders in den Blick genommen, aber die Reihe ist auch für andere Themenfelder offen.
Die Bände dieser Reihe versuchen, originelle Ideen und systematische Ansätze zu entwickeln, neue Lösungsversuche zu explorieren und allgemein souveränes und kritisches Denken zu fördern. Deshalb soll der Text bewusst für einen größeren Leserkreis verständlich sein, nicht nur für Fachleute.

The series aims to encourage a dialogue and a constructive debate between science and philosophy. It approaches this goal from both sides: On the one hand, the implications of scientific results and theories for philosophical thinking will be explored. On the other hand, a philosophical penetration and analysis of scientific problems, concepts and theories is intended. In this endeavour, different philosophical approaches are pursued. While the main focus is on the natural sciences and mathematics, the series is also open to other topics.
The volumes in this series seek to develop original ideas and systematic approaches, to explore new paths, and to generally promote independent and critical thinking. Therefore, the text should be understandable for a larger readership, not only for experts.

Cette série vise à encourager un dialogue et un débat constructif entre la science et la philosophie. Les perspectives de ces deux disciplines sont considérées tour à tour pour approcher cet objectif: d'une part, il s'agit d'explorer les implications des résultats scientifiques et théoriques pour la pensée philosophique. D'autre part, une considération philosophique et une analyse des problèmes, concepts et théories scientifiques sont réalisées. Différentes approches philosophiques sont abordées. Les sciences naturelles et les mathématiques sont particulièrement prises en compte, toutefois la série reste ouverte à d'autres sujets.
Les volumes de cette série tentent de développer des idées originales et des approches systématiques, d'explorer de nouvelles solutions et de promouvoir, d'une manière générale, un mode de pensée indépendant et un sens critique. Par conséquent, le texte devrait être compréhensible pour un plus grand nombre de lecteurs, et non pas seulement pour les experts.

Jürgen Jost • Guillermo Restrepo

The Evolution of Chemical Knowledge

A Formal Setting for Its Analysis

Springer

Jürgen Jost
Mathematik in den Naturwissenschaften
Max-Planck-Institut
Leipzig, Sachsen, Germany

Guillermo Restrepo
Mathematik in den Naturwissenschaften
Max-Planck-Institut
Leipzig, Sachsen, Germany

ISSN 2524-7549 ISSN 2524-7557 (electronic)
Wissenschaft und Philosophie – Science and Philosophy – Sciences et Philosophie
ISBN 978-3-031-10093-2 ISBN 978-3-031-10094-9 (eBook)
https://doi.org/10.1007/978-3-031-10094-9

Plannug/Lektorat: Iris Ruhmann

This Springer imprint is published by the registered company Springer Nature Switzerland AG
The registered company address is: Gewerbestrasse 11, 6330 Cham, Switzerland

Foreword

The present book outlines an impressive pioneering venture, the application of scientific, in particular mathematical methods to the understanding of the historical development of chemistry. Written by a chemist and a prominent mathematician, who master not only their disciplines but a wide range of historical and philosophical knowledge, it opens up a deeply innovative perspective on the history of science. When in the past quantitative methods entered historical discussions, they often came with a reductionist bias. Here they help understanding complex historical dynamics and pave the way for a transdisciplinary approach to the history of science, leading to a veritable "scientific history." This fascinating book provides both, a rich overview of the historical development of chemical space, in particular the spectacular growth of the number of known chemical substances, and an introduction to the sophisticated mathematical methods available to study this space and its expansion over time. The methods range from the theory of hypergraphs, via time series analysis and agent-based modelling, to computational semiotics and machine learning.

Chemistry is perhaps the most productive science. It prepares and explores an archipelago of different kinds of substances. Today, 118 chemical elements and more than 170 million chemical compounds are known, but the possibility of chemical reactions and production new compounds is by no means exhausted. Chemical space is continually expanding; the realm of organic compounds, in particular, seems to be endless. How to grasp this overwhelming chemical diversity? Is it possible to unravel the huge network of chemical reactions and to answer the question of how many compounds, or how many organic compounds, are possible in total? And why does the number of new substances between 1800 and 2015 grow with such a stupendous regularity, i.e. exponentially with a constant rate? Is it because the underlying epistemic network of chemical knowledge, embodied by new substances, and their reactions among each other is scalable? And if so, what does this tell us about the underlying material, social, and cognitive processes?

The authors approach these questions with a new perspective on the underlying historical dynamics, investigating the corresponding material, social and semiotic networks, and their interactions among each other, elaborating, in a most original way, an approach developed at the Max Planck Institute for the History of Science. The evolution of knowledge can indeed only be understood from such an encompassing perspective, taking into account semantic and semiotic structures such as those represented by chemical formulae understood as paper tools, but also the social processes underlying the formation of knowledge, as well as its material manifestation, in this case in the form of new chemical substances.

This book provides an unprecedented access to the challenges of chemical diversity using both computational and historical methods—in contrast to other ventures entirely ignoring the historical dimension. The book demonstrates compellingly that chemical productivity and chemical space depend not only on our understanding of nature—including knowledge about reactive functional groups, valency rules, thermodynamic laws—but also on historically varying technological and social conditions. Expansion of chemical space is as much a historical subject as is the evolution of chemical knowledge, and both of them are intertwined. The book thus bridges the sciences and the humanities in novel ways. It provides original answers to the question of chemical space, and the computational perspective it unfolds sheds new light on the history chemistry.

Berlin, Germany
August 2021

Jürgen Renn

Preface

In 2017 we began to discuss particular mathematical approaches to analyse the electronic information on the huge network of chemical reactions, which has been constructed by collecting and weaving together all reactions reported in the chemical literature from the eighteenth century until today. The temporal nature of the data feeding this network, combined with our historical and philosophical interests, gave shape to a strong historical facet of our project, which has led to some publications shedding light on the history of chemistry from a novel perspective, that is, from a data-driven and computational perspective.

Our early results already demonstrated the potential of a computational history of chemistry, namely detecting and analysing historical patterns at large scales. This goes beyond what can be accomplished by traditional historical means that rely on comparatively small datasets. Computational history of chemistry also allows for understanding and modelling the processes underlying historical events, which can be used to test accepted ideas and competing historical hypotheses.

After discussing with several colleagues, which included Michael Gordin, Jeffrey Johnson, Ursula Klein, Manfred Laubichler, Jürgen Renn, Alan Rocke, and Joachim Schummer, among others, we realised that we actually had not only information about chemical reactions and substances, which are part of the material aspects of chemistry, but also data on the social aspects of the chemical enterprise and that we could incorporate relevant information on the semiotics of chemistry as well. This expanded our local focus on the material aspects of chemistry to a more structured panorama for the *evolution of chemical knowledge*.

Although the history of chemistry has been among the most prolific subfields of the history of science (until 1950 history of chemistry was the second most prolific history of science, after mathematics. Currently it is the third, after mathematics and physics[1]), in general studies of the history of science, it has received less attention than that of physics or biology. Not to mention the still low interest for the philosophy of chemistry, in contrast to the large volumes of research on philosophy of physics, biology and even economy and medicine.[2] On the whole, by many historians and philosophers, chemistry is seen more as a field with a gradual accumulation of knowledge than sharp transitions. But this may, in fact, be an advantage for historical and philosophical studies, as we wish to argue. In contrast to other sciences, the data structure available in chemistry is much more coherent and uniform than in other fields. This presents a unique opportunity for modern computational approaches to detect and analyse large-scale patterns in the evolution of chemical knowledge.

But data is not enough to get knowledge and insight on the subject under study. Our early results on the historical evolution of the chemical reaction network indicated how historical data could be combined with formal and computational methods. Then, we began to think about treating the social and semiotic data of chemistry with similar methods and this book came about.

It is our central thesis that the moment is ripe to undertake a formal and computational study of the evolution of chemical knowledge. It is so because of three main reasons. First, data, computational power and formal methods are already available to undertake this enterprise. In our research, we have encountered and utilised very well-structured electronic datasets of information on the material aspects of chemistry. This information is also associated with relevant data about the social aspects of chemistry, which can be complemented with other social data available in non-digital form, but suitable of digitisation. Likewise, different historical sources of semiotic information are available, some of them already in digital form. This information, combined with the current—and still growing—computational power, ensures the storage and processing of information pertaining to the evolution of chemical knowledge. Still, data and computation do not provide a complete picture. Models are needed, not only because we humans think and communicate through them, but because computers require them as the basis for their algorithms. This third ingredient is also at our disposal, resulting from the different efforts to make sense of data and processes of different provenance, such as those of linguistics, complex systems, chemistry, economy, mathematics and computer sciences.

[1] Information retrieved from Ngrams from Google on 28th April 2021 by searching for "history of chemistry, history of physics, history of biology, history of mathematics" between 1800 and 2020 in the English corpus.

[2] Information retrieved from Ngrams from Google on 14th October 2021 by searching for "philosophy of chemistry, philosophy of physics, philosophy of biology, philosophy of economics, philosophy of medicine" between 1800 and 2020 in the English corpus.

This book, therefore, provides the basic elements for approaching the evolution of chemical knowledge from a formal and computational perspective. Moreover, as a referee pointed out, it also brings together "three diverse strands that since many decades stand quite separately in the history of chemistry and [... in the] history of science in general". We really hope the interplay of the three strands of our referee, namely those based on the social structures, the ideas and the practices, which we bring together as the social, semiotic and material ones, can lead to a unified and comprehensive picture of the evolution of chemistry, which can be actually used for better understanding the past, present and future of chemistry as a discipline. Such a picture is based on historical data interwoven with mathematical and computational tools and enriched by historical narratives and methods.

We begin our book by defining knowledge and chemical knowledge and by providing a formal setting for its analysis (Part I). In Part II, we discuss the data required to address the different aspects of chemical knowledge, and we provide a selection of formal and computational methods to gain insight about and from the available data. Although these parts may look, in principle, separated, we provide many connections between them by indicating which particular methods have been or could be used to address particular aspects and questions of the evolution of chemical knowledge.

Besides discussing some examples and results meeting our computational approach, readers will find several open questions and some possible ways to solve them. These questions not only pertain to the past of chemistry and to the historical events leading to the present chemistry, but also to possible pasts, that is, alternative histories of the discipline that were not realised. We emphasise that the use of computational approaches relying on curated data, combined with mathematical and statistical settings, allows for estimating future trends of the discipline. Hence, our research programme aims at casting light on the past, present and future of chemical knowledge.

As a book containing results and also several open questions, it is far from being a compendium of finalised research. On the contrary, we aim at motivating further research. The content presented here is not to be taken as a manual or as a guide to be followed strictly as we present it. It cannot be so for a simple reason, our approach requires interdisciplinary research at least among historians, chemists, physicists, computer scientists and mathematicians. We believe that a manual for such a complex interaction of disciplines with different research paradigms and methods to understand such a subtle and complex topic as chemical knowledge can hardly be encoded in a book.

The book is therefore written for an interdisciplinary audience, and therefore we have used, as much as possible, non-technical language accessible to readers of the sciences and the humanities at different levels of scholarship. Therefore, instead of describing historical, chemical, mathematical and computational details, we motivate our ideas with passages of the evolution of chemical knowledge, interpreted in a mathematical and computational framework. For the sake of clarity, we exemplify these ideas with particular past and present cases. For the technical details of our approaches, as well as for a deep exposition of our historical examples, we direct the reader to an ample and selected specialised literature.

Given the interdisciplinary character of our book and its non-technical language, we believe the book can be taken as a complement to more traditional material in undergraduate and graduate courses in chemistry, history of science and digital humanities.

While preparing the final version of this book, we organised an online meeting at the Max Planck Institute for Mathematics in the Sciences (22–23 March 2021) on "Computational approaches to the history of chemistry". This event helped us a lot to sharpen several of the ideas here presented and to know of other sources of information not considered so far by us. The meeting also contributed to support our belief that the moment is ripe to extend formal and computational studies of the evolution of chemical knowledge.

As it often happens, the content that we present here is not only a joint product of its authors. It is rather the result of multiple interactions and discussions with colleagues and students of different disciplines. Several of the ideas presented here have been motivated by discussions with Andrés Bernal, Gerd Blanke, Michael Gordin, Gerd Graßhoff, Evan Hepler-Smith, Roald Hoffmann, Ernst Homburg, Jeffrey Johnson, Ursula Klein, Manfred Laubichler, Duc Hoang Luu, Kärin Nickelsen, Carsten Reinhardt, Jürgen Renn, Alan Rocke, Jeff Seeman, Joachim Schummer and Peter F. Stadler, and many others. Very talented students such as Wilmer Leal and Eugenio J. Llanos have also significantly contributed with results and further discussions.

Leipzig, Germany
April 2021

Jürgen Jost
Guillermo Restrepo

Acknowledgements

We thank Joachim Schummer for inspiring philosophical and historical discussions about chemistry, Jeffrey Johnson for indicating suitable sources of information for the academy–industrial relationship in the nineteenth century, Peter F. Stadler for sharing his ideas on graph-rewrite rules and Ursula Klein for her comments upon a previous version of this document.

Contents

Part I

Chemical Knowledge as a Complex Dynamical System

Introduction 1

Abstract

We discuss the societal role of chemistry and its position among the sciences, which motivates the question of what chemical knowledge is. By defining knowledge, we characterise chemical knowledge as driven by social, semiotic and material aspects of the discipline. Finally, we present the advantages for the current and future practice of chemistry of analysing the evolution of chemical knowledge, as well as the social reaches of this analysis.

Chemical knowledge and its evolution[1] not only matters to chemists interested in their disciplinary past and future, it is of central scientific and societal importance, as chemistry shapes and creates the disposition of the world's resources, and lies at the border of science, industry, welfare and hazard [2]. Chemistry doubles its material output about each 16 years through publication of new substances[2] [4] and it is the most productive science in terms of number of publications (Fig. 1.1), only surpassed by Engineering.[3] The evolution of chemical knowledge is also of general societal interest because of its strong relationship with practical applications, especially with industry and technology, as documented by its

[1] Our topic is the evolution of chemical knowledge, which is different from the chemistry of (biological) evolution, often called evolutionary chemistry [1].

[2] On January 2022, Reaxys, one of the largest databases of chemical information, reports 57 million chemical reactions and more than 170 million substances [3].

[3] Derek John de Solla Price (1922–1983) already in the 1960s found chemistry as one of the disciplines of fastest growth ever in terms of publications and abstracts [5]. Joachim Schummer (1963–) confirmed it in the 1990s [6].

J. Jost, G. Restrepo, *The Evolution of Chemical Knowledge*, Wissenschaft und Philosophie – Science and Philosophy – Sciences et Philosophie,
https://doi.org/10.1007/978-3-031-10094-9_1

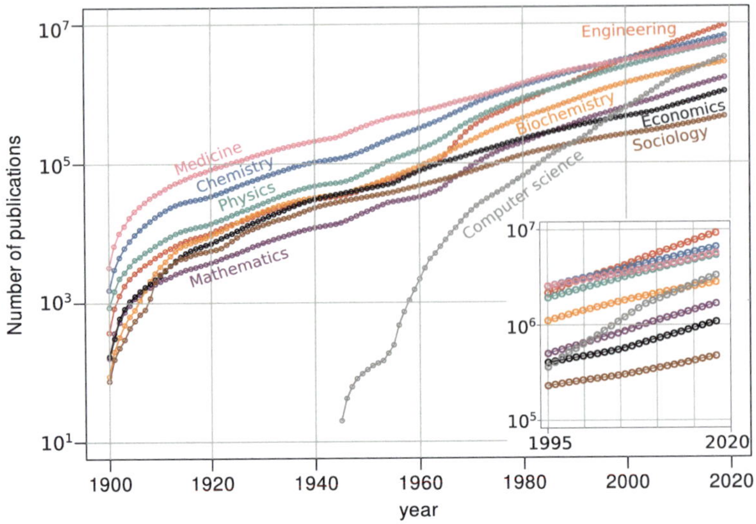

Fig. 1.1 Historical production of chemical literature. Cumulative output of articles of different disciplines for the period 1900–2019. Data retrieved from the Web of Science in November 2020 using the search tag "WC = discipline", corresponding to the Web of Science Category. Inset: zoom in of years 1995–2019

central position in patent applications [7].[4] Moreover, chemistry has been instrumental in the rise of biochemistry, molecular biology, material science and nanotechnology, to name but a few allied disciplines with their social and environmental impacts. Not surprisingly, Immanuel Kant (1724–1804) found chemistry as a paradigm for the method of critical philosophy, as he was astonished by the methods and logic of this science.[5] Kant's thoughts about chemistry come from the common methods of analysis, separation, purification, and synthesis shared by philosophy and chemistry [9]. His most famous statement about chemistry is, however, Kant's first position (1786) [9], which considers that chemistry could never be a science, as it did not proceed from a priori principles and therefore was not amenable to mathematisation. Once Kant learned about Antoine Laurent Lavoisier's (1743–1794) new chemistry, he asked his friend and colleague, the professor of medicine Karl Gottfried Hagen (1749–1829; in fact, Hagen founded the first pharmaceutical-chemical laboratory in Germany at the University of Königsberg, where both Kant and he were professors, thereby establishing the scientific discipline of pharmaceutical chemistry in Germany) to repeat Lavoisier's crucial experiments for him. Kant then changed his opinion about chemistry. Kant's post-Lavoisier position about chemistry, however, is only documented in his work from the 1790s, which was not

[4] The five scientific fields with the strongest connection to technological areas are materials science, electrical and electronic engineering, physics, chemistry and chemical engineering [7].

[5] For a systematic analysis of Kant's views on the nature and the structure of the sciences, see [8].

published during his lifetime and only became posthumously available, and only translated into English in 1993 in an abridged version [10], see [9, 11] for an analysis of its relevance for chemistry. Kant's conception of chemistry was dynamical, not material, based on forces instead of properties of matter, and consequently, he rejected the concept of an atom. For an analysis of Kant's theory of chemistry and its influence in Germany, see [12]. In the subsequent philosophy of German idealism, in particular in the systems of Schelling and Hegel, chemistry assumed a fundamental role. The question that arises is what actually knowledge and chemical knowledge is. This is elaborated in the next section.

1.1 Knowledge and Scientific Knowledge

Chemical knowledge is a special case of scientific knowledge, and the latter is part of human knowledge. Knowledge as such is an important, if not central, concept not only in the philosophy and history of science, but in a wide range of disciplines, ranging from philosophy and the theory of cognition to the social sciences. It is therefore a natural question what, if anything, those disciplines can contribute in providing conceptual insight for our topic, the history of chemistry. Therefore, we shall now embark upon an, admittedly rather selective, critical examination of the knowledge concepts of various disciplines from the perspective of this book.

From the perspective of computer science, knowledge expresses information, that is, reduction of uncertainty [13].[6] But information theory does not address the issue of the meaning of information, and this is an important issue for philosophy, see for instance [14]. And in biology or biochemistry, information as encoded in the physical properties and the geometry of the arrangement of its carrier, typically some macromolecule like DNA or RNA, is complemented by the structure of the environment [15].

In modern epistemology, knowledge came to be conceived as *justified true belief*.[7] The issue of truth itself is fundamental for philosophy. The criterion for truth is either the coincidence with some objective reality, or the coherence with a system of other propositions considered as true. While many scientists believe in the existence of some reality that is independent of the human mind, over the course of the history of science

---

[6] Interestingly, in that regard, chemistry not only accumulates information about its objects, but these objects may themselves encode information, like DNA or RNA molecules, and creating and manipulating chemical objects as information carriers, like memory alloys, may constitute an important future direction of chemistry.

[7] After Edmund Lee Gettier III (1927–2021) [16] came up with somewhat contrived examples where some proposition happened to be true, but was subjectively justified on the basis of false premises (for instance, in a conjunction "A or B", A was falsely believed, but B happened to be true, but was unknown to the subject, and so the conjunction was both subjectively justified and objectively true, but could not be considered as known to the subject), the field got into turmoil, with many attempts to justify, modify or extend the original definition. The details are not relevant here.

the content and the properties of such a reality underwent systematic changes. One may then consider prior states of science as incomplete or sometimes faulty insight into the objective reality as revealed and understood by current science, but such an attitude seems hardly adequate for a historical perspective, and it has been convincingly demolished by Thomas Samuel Kuhn (1922–1996) and others. But on the other hand, there seems to be scientific progress, and we think that we, as part of the contemporary scientific community, possess a better understanding of nature as earlier generations, as evidenced by more precise predictions of natural events and a superior technical control of natural processes. Therefore, a purely relativistic conception of knowledge does not seem to be appropriate either. In particular, we need to be somewhat critical of some implications of the sociology of knowledge. Karl Heinrich Marx (1818–1883) had already relegated knowledge to the superstructure built upon the more basic economic substructure. Knowledge thus becomes part of ideology. Following Karl Mannheim (1893–1947) [17], knowledge is a social construction, and in his wake, Peter Ludwig Berger (1929–2017) and Thomas Luckmann (1927–2016) [18] talked about the social construction of reality. While Mannheim himself exempted the natural sciences from his analysis, some of his followers also consider science simply as some form of ideology, on par, for instance, with religion, and thus as a tool employed to achieve social cohesion.

In view of the preceding, it might be useful to first distinguish different forms of knowledge. On one hand, there is explicit, declarative, conscious knowledge, a knowledge of *what* that can be explicitly formulated and formally analysed. On the other hand, there is implicit, procedural, tacit knowledge, a knowledge of *how* that can be routinely applied, but not necessarily or easily explicitly expressed. We all know how to formulate grammatically and stilistically correct sentences in our native language, but without proper training, we may not be able to formulate the underlying rules. Or, unless we suffer from prosopagnosia, we can easily identity the people that we know by their faces, but we have difficulties in describing in detail even the faces of our parents, children or spouses.

Moreover, there is individual and collective or distributed knowledge. No scientist is able to master all the knowledge of his or her scientific discipline, and it is one of the important and fundamental tasks of a scientific community to make that distributed knowledge available to its members. As we shall discuss in detail below, chemistry, in spite of being the largest of all scientific communities, has been particularly successful in that regard. And as we shall see, that comes from an interplay of the material system, with its conceptualised ingredients: the chemical substances and reactions and its often standardised equipment and procedures; of the semiotic system, with its standardised representations via chemical formulae, its systematic handbooks, and more recently, data repositories; and of the social system, with its university education and training, its societies and other forms of organisation.

In any case, chemical knowledge is constructive knowledge. While modern chemistry started to expand through the extraction of substances, most notably from plants and animals, it currently predominantly mixes artificial ingredients and produces artificial substances. The decisive step, instead of the extraction of some natural substance, often is the purification of the ingredients. Thus, chemists produce substances that have never before occurred in nature. Thus, it may not be appropriate to conceive of chemical knowledge as knowledge about some independent outside reality, but as procedural, although to a large extent explicit, knowledge about the production of novel substances and declarative knowledge about their chemical composition and properties. Knowledge becomes a self accumulating process.

We thus should approach the concept of knowledge from the perspective of the history of science. Following Jürgen Renn (1956–) [19], humans accumulate experiences and cognitively structure them through models, which leads to cognitive settings for perception, understanding and thinking. Cognitive models are also used for predicting further experiences and for communication [20–22]. Understanding arises as a structure of interconnected experiences allowing for abstractions, which result form the recognition of general structural patterns that become concepts or higher order structures of experiences. At this level of abstract structures, concepts are further interrelated leading to theories and to perception models that are used to acquire and accommodate further experiences.[8]

Knowledge, therefore is based on experiences, but the role of the observer is also crucial, in particular his/her capacity for processing information. Cognitive models develop and become more general and abstract as human brains mature from infancy to old age. Stable structural patterns appear often as new experiences are added, while transient ones are affected by these additions.[9] Which components of a model are or should be kept and which ones should be modified or abandoned in cases of conflicting evidence is studied intensively in psychology, see for instance [23], and philosophy, e.g. [24]. In this setting, knowledge is a dynamical cognitive model shaped by the unfolding of experiences and by brain maturity. In order to recognise stable and transient structures in the cognitive models constituting knowledge, these models need to be stored allowing for their sharing and transmission. Interestingly, storage and transmission is achieved by further cognitive models constituting languages, which require a material system to be preserved and spread, for instance through signs and text on surfaces or in digital form in magnetised materials [25].

Knowledge depends upon experiences, which are provided within an environment and therefore further depend on several factors, for instance social and semiotic ones. A person embedded in a political and economic system allowing for access to different sources of information, even of international provenance, is exposed to a totally different set of experiences than another person subjected to a controlled and restricted set of experiences

[8] A theory can accommodate given experiences and predict further ones, which when realised, strengthen the theory, and when not, enlarge the set of experiences and are used for developing alternative models.

[9] If we imagine a given set E of experiences, upon which we have identified a set V of relevant variables for describing E, a corresponding model M entails a network of variables in V. As E is modified, some relationships in M are strengthened while others disappear or may even appear. The former are the stable structural components of the model and the latter the transient ones.

driven by a particular political regime. Wars and pandemics constitute social conditions exposing people to atypical experiences as well. Language and other semiotic systems can both facilitate and hinder full appreciation of experiences. All in all, semiotic systems are cognitive models to perceive, store and communicate experiences and models built up upon them. Thus, knowledge depends on both external experiences and internal dynamic cognitive structures (models) which themselves not only develop over the lifetime of an individual, but are also created or shaped by social and cultural factors.

Scientific knowledge arises when experiences are addressed through further cognitive structures that emphasise some features of the experiences over others and when abstract structures such as theories are used to arrange the given experiences and to estimate further ones. The different scientific disciplines arise from this interplay of cognitive structures and type of experiences. Of particular interest for our discussion on chemical knowledge is the case of empirical knowledge. Schummer [26] uses the example of a coin which not only has material properties but also a symbolic or economic value that is largely independent of those material properties, to distinguish empirical knowledge from other kinds of knowledge.

1.2 Chemical Knowledge at the Centre of Material, Social and Semiotic Interactions

In the preceding section, we have discussed the issue of knowledge from the general perspective of the history of science. We shall now turn to chemistry proper, which prompt us to discuss what chemistry is about and what its knowledge encompasses.

At the core of chemistry lies the notion of material change, that is of change independent of Locke's primary qualities [27] such as extension (size), figure (shape), motion or rest, solidity (impenetrability) and number, which perfectly apply to the ontological level of bulk materials. Nevertheless, and especially in the twentieth century, the material entities of chemistry (or the former "bodies" of the discipline) suffered a change in scale and now figure or shape play an important role in chemical studies. This is evident, for instance, in the study of nanostructures, nanomaterials and in the growing number of organic and inorganic polymers.

Along the history of chemistry, some other material properties have been the focus of chemistry. In particular, properties resulting from the interaction of materials and also Locke' secondary qualities [27]. The latter refer to properties producing observer' sensations such as colour, taste, smell, sound or warmth and cold. The former are called *chemical properties* [26] and are central for chemistry because they result from the observation of behaviours of substances interacting with others in such a manner that a change of secondary qualities is observed.[10] This occurs, for example, when a colourless solution is mixed and shacked with a yelow solution and the resulting product is a colourless solution.[11] Actually, substance identity is given by a collection of properties, which have varied over the history from information on the sources providing the substance or material to properties related to theories about the microscopic constitution of matter (see Sect. 2.2.2 in Chap. 2).

Along the history of chemistry, a key cognitive structure devised to allow for studies of matter transformations is the concept of *chemical reaction*,[12] which encodes the temporality of matter transformation and is based upon the relationships among abstract objects: chemical substances.[13] The temporality of the chemical change informs about substrate(s) and its physical treatment, such as heating and stirring, going through its consumption and the appearance of the new chemical(s) and ends up in its separation, purification and characterisation. Chemical reactions relate *chemical substances*, actually pure chemical substances, which depend on the ontological level of description, for example of bulk materials or of quasi-molecular species.

---

[10] From a physical or physiological perspective, Locke's classification, which represents a key component of his empiricist philosophy, seems somewhat confused. His primary properties refer essentially to visual or haptic sensory experiences, but mix intrinsic properties like shape or solidity with external ones like position, or rather strangely, number. Colour is also visual, but, from a presentist account, as it reflects light absorption properties of objects, can give clues about their structure at the atomic scale. In that sense, observing the colour of an object might be the most basic spectroscopic method, that is, a method for inferring the internal structure and composition of an object through interactions with photons or other particles. Smell result from probing molecules through our senses, and therefore can give much information about chemical properties. Warmth and cold, or in a more recent terminology, temperature, is nowadays not considered as an intrinsic property of a chemical substance, but rather as an external condition that may facilitate or inhibit chemical reactions and change the phase of a substance. Water is a thirst quenching liquid only within a certain temperature range. When too hot, it causes the sensation of pain, which, while also caused by sensory neurons (the nociceptors), is not contained in Locke's classification. More generally, temperature changes the aggregate states of chemical substances, and for instance, the melting point is an important distinctive feature in chemistry. Thus, the characteristic property is not a temperature itself, but the response of a substance to temperature changes.

[11] This is actually an experimental setting to detect unsaturated compounds known as the bromine water test, where the colourless solution typically contains the substance whose unsaturation is to be tested and where the yellow solution is a mixture of bromine and water. This test, as well as several others [28] constituted powerful chemical tools in the nineteenth century to identify and characterise substances.

[12] A brief introduction to the historical evolution of the concept of chemical reaction and its representation is provided in pp. 156–157 of [29].

[13] Chemical reactions constitute the perfect setting for analysing chemical properties, as in a chemical reaction substances are brought together under controlled conditions to carefully study the effect of matter interaction.

The bulk material level approaches the concept of pure substance through experimental purification procedures, which typically involve phase transitions such as distillation, crystallisation and sublimation [26,30]. Here a substance is pure insofar as its characteristic properties do not change with a pool of purification or separation procedures. A typical characteristic property is the reaction to temperature [26]. Characteristic properties, as well as the purification procedures, are historical, as they depend on technological advances and theoretical settings, which change across time and finally result from negotiations and agreements within the chemical community. The ontology of quasi-molecular species, that is, of atoms and molecules, leads to pure substances defined as ensembles of quasi-molecular species holding a narrow range of energies. Such ensembles may undergo phase transitions, which connects them with the bulk ontological level. However, a single quasi-molecular species is not regarded as a pure substance because it cannot, by definition of phase transition, undergo this kind of process. The quasi-molecular definition depends on the range of energies characterising the ensemble. For very narrow energy levels, several quasi-molecular species sitting on a large energetic range are rule out, therefore increasing the amount of pure substances.[14]

Currently, the quasi-molecular description level for substances is strongly concentrated on the different geometric possibilities of quasi-molecular species. This is the case of stereoisomers and enantiomers, whose dramatic importance became evident through the case of thalidomide in the 1950s [31], which brought to the fore of the social arena the terrible health effects of regarding as chemically and physiologically equivalent two molecules with slight geometric differences. Actually, the difference of being not superimposable mirror images of each other, i.e., having the same constitution and patterns of chemical bonds but a different spatial arrangement of those. At a larger molecular scale, that is of macromolecules, geometry is also central. The function of a protein in a cell depends on the geometric shape it assumes as the result of a folding process, whose precise details are still not completely understood. Even when some of its constituents, the amino acids, are changed as a result of genetic mutations, this does not necessarily affect the function, as long as the geometric structure is preserved. And presumably, some molecules in a cell, for instance certain types of RNA, can act as scaffolds for bringing different types of proteins together so that they can interact. The DNA in a cell contains many segments that were too hastily disregarded as "junk DNA," but which, in cooperation with various proteins, seem to play an important role in the spatial organisation of that giant molecule. That spatial organisation of DNA and the formation of special spatial motifs in RNA is needed for the selective expression of specific sets of genes according to cell type and condition, see [32]. The current vibrant field of material science, at the border of chemistry, physics and engineering, entirely depends on geometrical arrangements of atoms and quasi-molecular species, where small geometrical variations may lead to materials with different macroscopic properties, such as hardness and electrical and

[14] Further discussion on the semiotic evolution of the concept of substance is found in Chap. 3.

thermal conductivity [33]. Also, some materials can switch between different geometric configurations depending on external parameters, and memory materials can be induced to return to their original shapes [34].

Despite the important role of substances for chemistry, this science goes far beyond the identity of its substances. Chemistry is mainly concerned with substance transformation and it is definitely not interested in substances as individual objects. This explains why, as mentioned before, Locke's [27] primary qualities, are not that central in chemistry.

A chemist may distinguish between the intrinsic properties of chemical substances, that is, the dispositions and the availabilities for reactions, and their extrinsic uses as materials, pharmaceuticals, fertilisers, dyes, detergents or whatever. And as already emphasised, chemical knowledge is not about individual objects or events, but about invariant properties and reproducible processes.[15]

By observing and acting upon matter transformations, chemists have come up with further concepts aiming at better understanding and forecasting those transformations. Some of these concepts involve substance and reaction classifications, which have led to further concepts such as periodic systems and functional groups and to affinity and bonding theories that go hand in hand with the appearance of the powerful semiotic concept of molecular structure, today extended to the structure formed by atoms and their relationships that includes molecules, as well as inorganic and organic polymers and alloys. Matter transformation has provided the experimental evidence to devise theories upon the speed of those transformations and the conditions for them to take place. They involve chemical kinetics, thermodynamics, quantum chemistry and reaction mechanism studies (Fig. 1.2).

But from our perspective, chemical knowledge goes beyond a narrow semiotic realm conceptualising some specific relevant material properties and dispositions. Clearly, synthesis and analysis, nowadays assisted by spectroscopic methods, constitute key components of chemical knowledge. Procedural knowledge is also necessary. That is, how to carry out specific experiments, including the required preparatory steps like purification and the current computer simulations or the search in chemical databases. Knowledge from other fields, in former times from metallurgy and pharmacy and today from biology, physics and mathematics, is needed. The social component is likewise important. Chemists need to know about the economic availability of possible scarce ingredients and appropriate laboratory equipment. They also need to know, for example,

[15] There are some relevant issues here. A chemical substance is not an individual molecule. So, a chemist is not interested in a specific molecule of H_2O, but in the properties of water, at different temperatures and in different forms of aggregation. It is not always so easy, however, to draw the line between a material or a substance on one hand and a specific molecule contained in it or representing it. An interesting chemo-mathematical question spanning centuries of theoretical and experimental inquiry is determining the minimal material sample condensing the material properties of the bulk material. These minimal entities include molecules, unit cells and proportion of components in new materials.

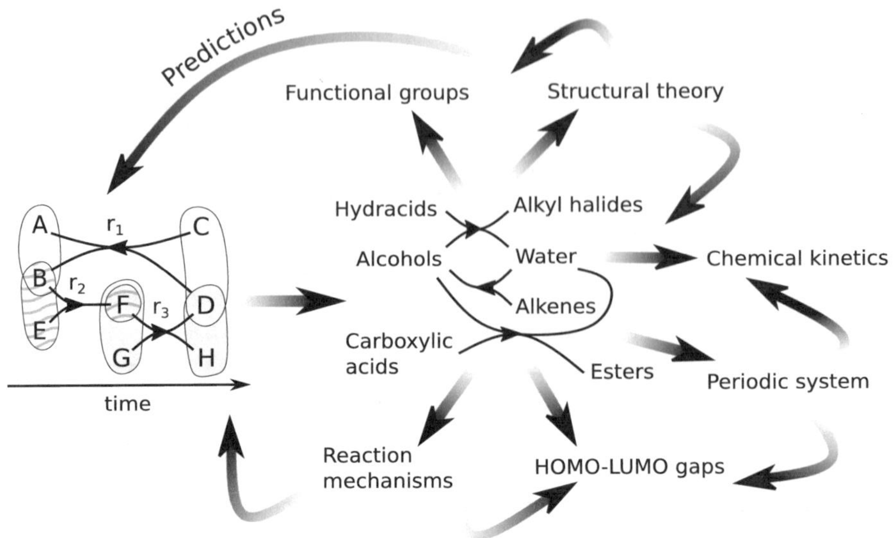

Fig. 1.2 Abstraction of chemical knowledge and its material ties. Left network represents a reaction network made of three reactions: r_1: C + D → A + B, r_2: B + E → F and r_3: F + G → D + H. Each reaction is represented as a hypergraph (see Sect. 6.1) made of two same-black/white-pattern subsets, where one subset gathers together the substrates of the reaction and the other the products. The reaction network evolves across time by the addition of substances and reactions, which leads to a system of classification of substances forming a network of classes (central network). This network produces concepts and theories in chemistry, which are further used and intertwined to make predictions

how and where to obtain research funding, how to recruit laboratory assistants and technicians, or how to spread their findings taking into account the social conditions and the available communication channels. Thus, chemical knowledge, besides the material realm, also depends upon social and semiotic aspects.[16]

[16] This idea has been already discussed in the history and philosophy of chemistry, as evident for instance in Leicester's book [35] from the mid twentieth century, when he states:

> It is clear that the full story of such [historical] developments involves not only the personalities and intellects of the scientists themselves, but also the social and economic conditions which surround them and the philosophical ideas to which they are exposed.

There have been several academic movements advocating a history and philosophy of chemistry in practice, where theory and the world are considered simultaneously. Examples include the International Society for the Philosophy of Chemistry [36] and the Society for Philosophy of Science in Practice [37]. For more historical insight on the development of this setting for the history of science, see [19] and footnote 20 in this chapter.

Generation of new knowledge requires careful study, research or application of skills, and this may naturally lead to professionalisation. While at the beginning of chemistry, metallurgists, alchemists, pharmacists, or other individuals that had some skill and experience in the extraction and transformation of substances could make important contributions, nowadays a PhD in chemistry or in a related field is required to get access to the research infrastructure on which chemical research depends, and given the current state of the field, outsiders no longer have a chance to make contributions. Of course, one can argue to what extent this is really required on scientific grounds, or whether the profession has simply created certain barriers to strictly regulate and control access to the field, not dissimilar to the guilds in medieval times. It is not our purpose to enter such a discussion.

While in the early phases of the development of a particular science, there may have been some individuals, like the founding fathers of a discipline or some systematic scholars, that were able to accumulate and master more or less the entire knowledge of their discipline, subsequently, the acquisition, storage and transmission of knowledge inevitably became a collective process, and in turn, the coherence of some body of knowledge, of its social organisation and of its material representation might delineate a particular scientific discipline during a particular period.[17] Across time, knowledge dynamically changes due to new experiences, cognitive reorganisation, social evolution and the developments of material and semiotic systems. Of course, all these processes are intertwined, and it is a central task for any science to understand the contributions and the interdependencies of these factors.

From the beginnings of modern science, progress depended in an essential way on the availability of mechanical precision measurement devices, that is, on the skills of craftsmen. These skills were usually originally developed for other cultural or economic purposes. Nowadays, experimental science requires highly specialised and typically very expensive equipment. This implies not only that new discoveries depend on external financial support, and thus on the economic means of the society in which research takes place, but also that often new results can no longer be confirmed by independent research groups. When the director of CERN's Large Hadron Collider (Europe) announces the discovery of a new elementary particle, or when scientists at the Lawrence Livermore National Laboratory (USA), the Joint Institute for Nuclear Research (Russia), the Helmholtz Center for Heavy Ion Research (Germany) or the RIKEN Nishina Center for Accelerator-based Science (Japan) announce the synthesis of a new element, we need to rely either on trusting the careful internal scrutiny or have to evaluate how plausible the discovery is from the existing theoretical frame or whether it gives rise to any contradictions. But of course, such new discoveries may also require changes of the underlying theory. Moreover, many such discoveries are based on rather indirect evidence [38].

[17] Important preservers of alchemical knowledge were, for instance, Byzantine alchemists, who laboriously copied and commented upon Greek sources when the Western Roman Empire was fading away [35]. Others are discussed in Chap. 4.

In order to be accepted by the community, new chemical species no longer need to be stored in bottles and flasks, but only leave clear, but usually rather indirect traces of their existence, like spectra or X-ray signals. New superheavy elements, for instance, need to exist (whatever that means) only for extremely short periods of time, like 10^{-14} seconds [39]. Thus, the criteria for accepting the discovery of a new substance have to be carefully negotiated and agreed upon by the scientific community.[18] This is in stark contrast to a theoretical formula or a mathematical proof that can, at least in principle or according to the scientific narrative, be checked by any competent expert. In experimental science, the tools for the confirmation or perhaps also the falsification of an announced discovery are available at best to a small circle of usually closely connected or intensively competing laboratories.

One and the same set of scientific data can be organised differently. The trajectories of the celestial bodies, for instance, can be arranged in a geocentric or a heliocentric frame.[19] Thus, knowledge depends not only on the underlying data, but also in an essential manner on how they are scientifically arranged and structured. In general, it even depends on the scientific framework and aim what counts as a datum. One can argue much about the constructive aspect of cognition in general and science in particular (e.g. [41]).

As one of the referees of this book pointed out, scientific disciplines, including chemistry, become crucial social and epistemic ordering systems [42–44], where science as a cultural practice is deeply intertwined with other human practices, such as social, political and even aesthetic ones.

It is our thesis that chemical knowledge emerges from the interaction of social, semiotic and material aspects. This brings forward a unification of the three seemingly different traditions/schools in history of science, each one emphasising either the social, or the

[18] See footnote 15 in Chap. 4.

[19] Brock provides an alchemical/chemical example of different interpretations of the same experimental result:

> If a mixture of nitric acid, mercury, and silver are allowed to stand for a couple of months, the mix 'vegetates' into a tree-like form of silver called Diana's Tree. But whereas alchemists had seen this tree as evidence of a vital spirit and a key step en route to the philosophers' stone, French chemists discussed it under the properties of silver and as an example of mechanical crystal formation [40].

A further example of different interpretations of the same experimental fact is provided by Hennig Brand (c. 1630–c. 1692), who believed that fire was composed of small particles in rapid motion [35]. As reported by Henry Marshall Leicester (1906–1991) [35]:

> [Brand] sealed various metals in glass flasks and heated strongly to convert them to their calces (oxides). He then opened the flasks and weighed the product. The increase in weight was attributed to fire particles that had penetrated the glass and combined with the metal.

Today, the increase in weight is associated to the addition of oxygen to the metal to form oxides.

ideas, or the practice of scientific disciplines [19] (pp. 55–57).[20] This book therefore offers a unified framework for the case of the history of chemistry and the evolution of chemical knowledge. But it goes beyond that unifying framework by showing how it can be put into a mathematical setting and how such a setting can be used to understand and even foresee the unfolding of chemical knowledge. Our formal setting (discussed in detail in the coming chapters) should not be merely regarded as an abstraction of a fragmentary historical discussion that went over several decades, concerning not only chemistry, but the whole edifice of science. Our approach goes beyond formalisation, it actually uses it for the benefit of chemistry, its knowledge and the historical studies of its evolution. In the following chapters we provide several application examples of our setting and indicate new research avenues where huge corpora of historical information are computationally integrated into our framework to shed light on the past, present and future reaches of chemical knowledge.

Moreover, as we discuss in this book, mathematics is mature enough to assist the approaches here presented, which when combined with the digitisation efforts spanning the different aspects of chemical knowledge, ranging from social and semiotic, to material ones, make this moment in history the right one for undertaking a data-driven study of the historical unfolding of chemical knowledge. In the following chapters we present a research programme to undertake this enterprise, where we discuss the different sorts of available data, as well as the computational and formal approaches to treat this corpus and to shed light on the evolution of chemical knowledge. This book, all in all, unifies the different historical stances about chemical knowledge through formal methods, computational capacity and large corpora of historical information.

Before proceeding, we motivate the formal study of chemical knowledge by highlighting particular benefits it provides for chemistry.

---

[20] Renn argues that this approach to the history of science from three different perspectives or strands was motivated by specialisation, ideological divisions, World Wars, the Holocaust and the Cold War [19]. He, nevertheless, highlights important 1930s works addressing the history of science from a more holistic perspective integrating the three strands. A methodological stance that was forgotten in the practice of the history of science [19]. The 1930s examples discussed by Renn on the basis of an analysis by Gideon Freudenthal and Peter McLaughlin (1956–) are the so-called "Hessen-Grossmann theses," based on the pioneering works by Boris Mikhailovich Hessen (1893–1936) [45] and Henryk Grossmann (1881–1950) [46, 47]. These theses are: "First, that technology opens cognitive horizons for science, explaining the material conditions under which certain abstractions become possible. Second, that technology also limits the horizons of science, explaining why certain abstractions are not possible under given conditions. And third, that other social factors—including ideologies, beliefs, and political and philosophical theories—can affect science." [19] (p. 55). Another example of a holistic stance for the history of science of the 1930s discussed by Renn is Ludwik Fleck's (1896–1961) seminal book [48], where the Polish scientist introduced concepts such as "thought collective" and "thought style."

1.3 Chemistry's Benefits from Studying the Evolution of its Knowledge

As discovering new substances is at the core of chemistry [26], analysing the historical driving forces of this process may lead to detecting the conditions suitable for speeding up the exploration of the *chemical space*, which spans all known chemical species endowed with a notion of relatedness [4, 49].[21] Although chemistry has had a stable 4.4% annual growth rate of new chemicals from around 1800 until today (Fig. 1.3), there have been periods of rapid exploration as that between 1870 and 1910, driven by the rapid growth of organic chemistry [4], a period of "genesis of land and sea from chaos," as once Rolf Huisgen (1920–2020) put it [53]. During this period chemistry became the largest discipline in terms of number of practitioners [54, 55] as a consequence, among others, of a successful academe-industry alliance [56].[22]

Under which conditions, if any, can one—temporarily or permanently—go beyond the historical 4.4%? Can this be possible at least for certain subregions of chemical space, and are there good historical examples? Finding these conditions for particular regions of the chemical space, such as the Lipinski space of oral-drug like substances, would have remarkable societal consequences.[23] Can we learn from the prolific 1870–1910 period? A time where the chemical social structure was appropriately adjusted to the complexities of the chemical space. Or are there systematic reasons why the observed growth rate

[21] To better understand the chemical space, one may think of a Euclidean space, where elements of a set are assigned a metric allowing measuring distances among them and endowing them with a notion of nearness. The chemical space, in turn, entails a set of substances endowed with a relatedness notion given by synthetic paths connecting substances or by molecular resemblance [49]. The first relation leads to a network of chemical reactions and the second to the study of the molecular structure. Interestingly, the current surge of interest for chemical reaction networks has not always been the case. This has been motivated by the digitisation of reaction information, which grew in the 1980s, 1990s and which is currently analysed using machine learning tools [50], for example for the optimal design of synthesis plans. In the 1970s and the years to come, most of the interest was in a chemical space where the notion of nearness was molecular resemblance [51]. Several mathematical and computational methods were developed to quantify such a similarity, which eventually led to the paradigm of quantitative structure-activity relationships models, of widespread use today in medicinal and environmental chemistry [52].

[22] Other factors contributing to the popularisation of chemistry and its social nineteenth-century appeal include the influence of textbooks and the health problems chemistry solved [31]. Nevertheless, as we discuss in further sections, chemistry has not always been so popular. Its social image has varied from pure charlatanry in alchemical times to be the central science as well as one endangering our planet and civilisation.

[23] This space was defined by Christopher A. Lipinski and collaborators in a seminal 1997 publication [57], where they found that most of the oral-drugs fall in a particular region of a space of substance properties. Chemicals in this region weigh less than 500 daltons, hold an octanol-water partition coefficient lower than 5 and have no more than 5 hydrogen bond donors and no more than 10 hydrogen bond acceptors.

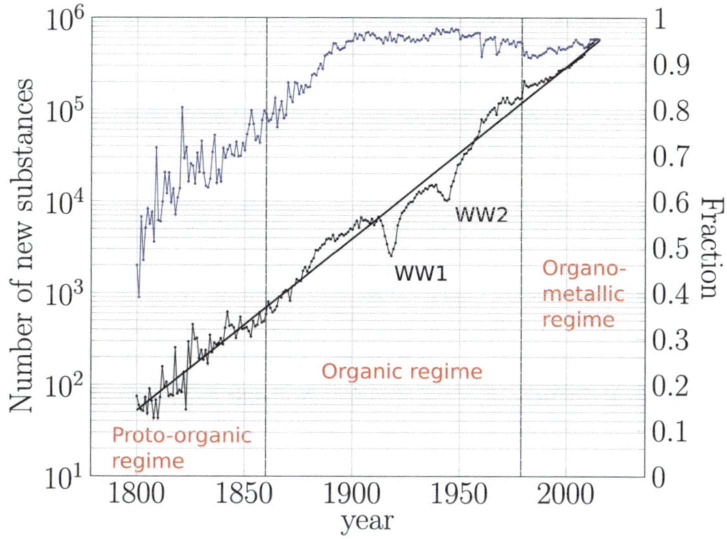

Fig. 1.3 The expanding chemical space. Annual growth of number of new substances between 1800–2015 (bottom plot, left axis). The effects of the World Wars (WWs) are indicated. The three statistical regimes resulting from the annual variability of the number of new substances are shown, with transitions occurring in 1860 and in 1980 (dotted vertical lines). The proto-organic regime is characterised by a relative high proportion of inorganic chemicals, when contrasted with organic ones. The organic regime is strongly biased toward organic compounds made of carbon, hydrogen, nitrogen and oxygen combinations and the organo-metallic regime depicts an initial rise of these compounds followed by a surge of bio-organic substances [4]. The exponential growth is indicated as a straight line, with equation: $s_t = 51.85\,e^{0.04324(t-1800)}$, where s_t is the number of new substances at year t. Fraction of new synthesised compounds to the total of new ones (upper plot, right axis). Adapted from Fig. 1 in [4]

of 4.4% is so stable, for instance a systematic delay in the diffusion of knowledge, the time it takes to train a new generation of chemists that can then apply new methods more systematically, a certain inertia of the scientific system, or extraneous factors related to economic growth, demographic developments or societal demands? In the latter case in particular, a similar limit should also be observed for sciences other than chemistry. It may also be the case that a growth rate becomes self-regulating as for example funding agencies and other institutions might come to expect a particular growth rate and plan their budgets accordingly. Some of these issues will emerge in our subsequent discussion.

Chemists expand the chemical space mainly through chemical reactions,[24] which they have historically tried to improve in terms of yields and more recently by minimising

[24] New chemicals also come from extractions from natural products. However, we have shown [4] that the role of synthesis, traditionally thought as important only after 1828 Wöhler' synthesis of urea, was a relevant source of new chemicals from the early years of the nineteenth century (Fig. 1.3).

the environmental hazard.[25] The analysis of the giant network of chemical reactions reported over the years in the literature indicates whether synthetic design is becoming more efficient, for instance by reducing the number of steps to chemical targets, or by avoiding the use of toxic solvents. The analysis of the network also indicates whether such anthropogenic wiring of substances through reactions is becoming more similar or not to metabolic networks. These networks make large use of their metabolites in a closed fashion such that reaction products become substrates of other reactions and substrates are produced by several reactions [59].[26] Is green chemistry actually making the two networks closer to each other?[27]

Analysing the evolution of chemical knowledge allows for determining its stability and instability under perturbations. For example, the COVID-19 pandemic, which has destabilised knowledge production [63] by affecting wet-lab experimentation and traditional collaboration dynamics; or gender gaps [64] and diversity issues in chemistry [65,66], with their impacts on the advancement of chemical knowledge. Other perturbations involve wars, which besides destabilising the research funding, reducing the number of active chemists and disrupting the educational infrastructure, surprisingly have not destabilised chemistry annual output of new substances [4]. After the expected drop in war times, production followed a catching-up recovery phenomenon that contrasts with typical production delays of other sorts, such as publication of abstracts in other disciplines [5] (Fig. 1.3).

Central for evolving systems is their capacity for innovation, that is novelty with lasting influence. For example an innovative reaction is a chemical transformation rule that is

[25] Reducing the environmental hazard involves also avoiding the synthesis of undesirable side products. Although increasing the yield of a target product is always desired, sometimes low yields of targets may be more important than high yields associated with side products that are difficult to remove from the target or highly toxic [58].

[26] The idea of cyclic reaction paths is central for metabolic processes. These reactions have been also deeply studied from a historical perspective. For instance Kärin Nickelsen (1972–) and Gerd Graßhoff (1957–) used a computational approach to reconstruct the discovery of the first cyclic reaction producing an important metabolite, namely urea [60,61].

[27] The comparison between the anthropogenic network of chemical reactions and the metabolic one entails a geometric idea. Can we measure geometric proxies of these two networks and compare them? We believe we can do it and we elaborate more on this point in Sect. 6.1. At any rate, the idea of geometric objects underlying chemistry has been part of the evolution of chemical knowledge and it is especially evident in the eighteenth-century efforts of the Enlightenment, where the very idea of "system" entailed a geometric meaning [62] (Chap. 3). Furthermore, the question of whether anthropogenic chemistry is similar to metabolic chemistry may be read as whether the first *ought* to be similar to the second, which boils down to the aim of chemistry and the ethics of the chemical activity (see Sect. 2.2.1). Nevertheless, the idea of "best geometries" has been around ever since the Enlightenment, for instance in the botanical works by Michel Adanson (1727–1806) [62] (Chap. 3).

new according to the chemical knowledge of a particular time and that is used by the chemical community for a certain period of time,[28] for example the Suzuki-Miyaura coupling, first reported in 1981 and of wide use ever since.[29] Studying the evolution of chemical knowledge allows for determining how innovation occurs, whether in a stochastic manner or rather whether it is driven by internal or external factors of chemistry, or by the interaction of both. Analysing the evolution of chemistry further allows for quantifying the acceptance time of chemical breakthroughs and whether there is a critical number of supporters for an idea/technology to be caught up.[30]

Understanding the evolution of chemical knowledge facilitates the detection of the conditions leading to knowledge crises such as the semiotic one of the years before 1860, a period when the deluge of organic substances challenged the nomenclature and the combinatorial capacities of the semiotics of chemical formulae. This crisis motivated the introduction of molecular structures as semiotic icons and paper tools [71], which besides enriching the theoretical repertoire of chemistry allowed for devising and synthesising substances difficult to foresee with the semiotics of empirical formulae and which, ultimately, led to the unusual rapid discovery of new chemicals between 1870 and 1919 (Fig. 1.3).[31] Another crisis is the current one, where traditional chemistry communication channels cannot cope with the rapid output of scientific results. This has led to using new forms of communication such as preprints, blogs, media posts and open-notebook initiatives, likewise it has favoured the mushrooming of predatory journals [72].[32] This has been particularly evident through the new dynamics and publication speeds caused by

---

[28] Some of these reactions are discussed in [67, 68].

[29] In [69] it is argued that innovation for the chemical system entails devising new reactions and substances that fall out of the scope of the available chemical knowledge, including known reactions.

[30] A related subject studied by sociologists of science and scientometricians is the case of sleeping beauties [70]. That is, of influential subjects of a discipline requiring a certain time (sleeping time) to be accepted by a community. The sleeping beauty entails a "kissing prince" embodying the conditions around the acceptance of the sleeping subject. Chemistry is after physics the scientific discipline with most sleeping beauties [70]. Examples in chemistry include Freundlich's 1906 paper on sorption processes awakened in 2002 as a suitable method to assess the accuracy of the fit of an isotherm model to experimental equilibrium data, or Hummers and Offeman's 1958 preparation of graphitic oxide revived in 2007 for the production of graphene [70].

[31] Eighteenth-century affinity tables (discussed in Chap. 3) were also important paper tools to expand the chemical space. An instance of this is given by Torbern Bergman (1735–1784), who in 1775, based on the affinity table formulated in 1718 by Etienne François Geoffroy (1672–1731), published tables for 59 substances undergoing reactions in solution and by fusion. He estimated that determining all the reactions of the substances in his table would require more than 30,000 experiments [35]. Incidentally, the qualification of affinities turned into quantifications through the ideas and methods of the eighteenth century, which led to determine the amounts of bases required to neutralise acids and, in the end, to equivalent weights, which were in turn instrumental for the atomistic ideas of the nineteenth century leading to the formulation of the molecular structural theory.

[32] A detailed analysis of the current publication system is found in [73].

the COVID-19 pandemic [74,75]. A particular semiotic crisis comes from the use of string molecular representations such as InChIs,[33] which fail to encode non-organic molecules [77,78] that despite being a tiny portion of the known chemical space, are instrumental for the exploration of the overpopulated organic region of the space [79]. This crisis, therefore, hindered and still hinders the storage and spread of inorganic chemistry knowledge.

As studying the evolution of chemical knowledge involves gauging the mechanisms leading to the emergence of macroscopic effects observed over the history of chemistry, this entails modelling and estimating the future of chemical knowledge. This forecasting is inherently stochastic, providing probabilistic results, such as determining the likelihood of having an innovative chemical reaction given certain material, social and semiotic conditions. This contrasts with deterministic predictions such as when robots will replace chemists or whether solar energy could efficiently split water into hydrogen and oxygen.[34] Through estimations, it can be better understood how the social structures influence chemical knowledge and what the possible consequences may be, which in turn suggests the actions the community needs to take to foster or to avoid such consequences. For instance, one could determine the social conditions leading to chemical weapons and the required adjustments to reduce this threat.[35] Or how political discourses, polarised opinions and even scientific evidence indicating the risks posed by some chemicals, may affect the regions of the chemical space to explore or to ban.[36]

Estimating entails constructing a probability distribution on the basis of a training set that is a random and incomplete representation of the systems constituting chemical knowledge. We have shown the difficulties in estimating the number of new substances based on historical records of the annual output of new chemicals [4]. The question that arises is what additional information is needed to obtain accurate estimates. We have

[33] International Chemical Identifiers (InChIs) are strings used to encode as much as possible information about chemical substances. InChI philosophy is based on the hierarchical incorporation of several of the modern semiotic layers of substance identification, from molecular structures to structures isotopically labelled, or at the level of stereochemistry [76]. Interestingly, InChIs were motivated by the increasing complexity of molecular structures, which make conventional naming procedures inconvenient [76].

[34] Murray Gell-Mann (1929–2019) provides a clear example in [80] of how historical data convey reliable estimations even if the workings of the estimated historical process are unknown. In Gell-Mann's example, the process is "the average distance covered in a day of travel using a vehicle, taking into account the mix of transport modes," which include travel on horseback or by horse-drawn coach, as well as railway, automobiles and airplanes. The result is that this average distance increased exponentially over the last two centuries with a stable growth rate [80].

[35] An interesting book on the history of chemical warfare is [81].

[36] In a forthcoming paper, we show how Donald John Trump's (1946–) tweets of March 2020 motivated research on hydroxychloroquine and chloroquine. An instance of banned regions of the space, which have therefore motivated the exploration of others, is the 1974 evidence that chlorofluorocarbons deplete the stratospheric ozone layer [82]. Likewise, per- and polyfluoroalkyls have been linked to cancer, immune system and developmental problems, among other health problems. This has triggered the ban on their production [83].

recently found that the annual output of new substances follows a memory process, which is presumably related to the conservatism in chemistry [4]. This result suggests that the production of new chemicals can be modelled as a dynamical system whose evolution depends on the system own history through the iterated application of an underlying dynamical rule. The challenge now is determining such a rule.[37] Likewise, it is important to undertake similar studies to estimate social and semiotic aspects of chemistry.

Estimations also involve determining the suitable reaction conditions to discover strange substances of presumably unexpected properties, which entails going beyond the synthesis planning to produce substances that somehow are expected according to the available chemical knowledge. Vitamin B12 [84] was a challenge not because of its associated exotic molecular structure, but rather because of the difficulty it posed to the synthetic knowledge of the 1970s. Chemists need to devise substances impossible to foresee with the chemical knowledge of their times and to link those chemicals to suitable reaction conditions leading to them, even if those conditions are outside the known possibilities. Therefore, estimating reaction conditions not only involves substances and reactions, but understanding the interplay between known conditions, chemical instrumentation and available chemical theories.

Besides estimations, philosophers and historians of science have commented upon the marked methodological shift that is underway in their disciplines, namely the incorporation of computational models as a tool for understanding [85–87]. Traditionally, explanations result from subsuming phenomena under laws of nature. Computational models open the possibility of exploring the development of interacting objects under changing conditions, which may determine the objects in themselves. As stated in [87], this setting removes the boundary between narratives and explanations and it is highlighted as a convenient approach to attain both narrative and explanatory discourses in a single framework.[38] Hence, computational models have a double function in history and philosophy of science, as tools for estimating the evolution of knowledge and to narrate/understand such an evolution.

Another facet of modelling the evolution of chemical knowledge is that it allows us to test historical hypotheses. Often historical discourses assume causal models leading to the target historical facts. Such discourses can be proposed, discussed and disputed, but because they depend on qualitative and subjective selections, weightings and interpretations of facts, they are difficult to verify or refute. Take the case of the formulation

[37] This rule may be gauged by analysing historical records of production of chemicals, which leads to moot growth models for the expansion of the chemical space. This is part of an ongoing project in our research group, where we use hypergraphs to model the network of chemical reactions. Alternative approaches may be based on agent-based models. See Chap. 6.

[38] In [87] this approach is exemplified as: "Real snowflakes, in contrast to their idealized image, are non-symmetrical and variable, and these features can be accounted for by simulating the growth of snowflakes under a variety of conditions. Such diversity can be expounded best by tracing many individual trajectories under different initial and boundary conditions."

of the periodic system. Was it the result of the chemical genius of Julius Lothar Meyer (1830–1895) and Dmitri Ivanovich Mendeleev (1834–1907)?[39] Was it a consequence of the known chemical space by the mid nineteenth century? It has been claimed [88, 89] that the 1860s were the ripe moment to devise the system. Moreover, it is recognised by historians that the standardisation of atomic weights in 1860 triggered the formulation of several periodic systems in the 1860s [90, 91]. In [92] we found that the chemical space of the 1840s already contained the salient features of the periodic system of the 1860s, which leads to ponder the social and semiotic reasons delaying its formulation more than two decades. The same study shows that by computationally adjusting the chemical space to the different nineteenth-century competing sets of atomic weights, Leopold Gmelin's (1788–1853) atomic weights of 1843 gauged the available periodic system of the time, as well as most of the structure of the system of the 1860s. The study allowed us to play historical games of the sort "what would have happened if," as we could figure out the possible periodic systems that, for instance Dalton with his atomic weights of 1810 and the chemical space at that time could have obtained.[40]

The same methods used to analyse the rise of the periodic system can be used to analyse its evolution and to estimate its future [39, 49].[41] This, likewise, makes it possible to test chemical theories with an augmented set of facts, which contrasts with the few data that came along with the theories in the past. This has been the case, for example, in a recent analysis of Pauling's rules, initially published in 1929 [107], about crystal structures [108]. These five empirical rules became the cornerstone of solid-sate chemistry and have been associated with the expansion of the chemical space of ionic compounds [108]. In [108] it was found that the rules only apply to less than 15% of the ionic compounds. Hence, modelling the evolution of chemical knowledge goes beyond the assessment of historical narratives, it embraces estimating the future of chemical knowledge and the validity of its theories.[42]

[39] Here, aspects of the chemical community play a major role, for example with the nationalistic overtones justifying "a" discoverer of the system, as discussed in [39].

[40] The study in [92] was carried out for the atomic weights and available chemical spaces of Dalton [93], Thomson [94], Berzelius ([95–98]), Gmelin [99], Lenßen [100], Meyer ([101], Rex [102], Meyer [103]), Odling [104], Hinrichs [105] and Mendeleev [106].

[41] Another example of testing competing historical hypotheses is the work by Graßhoff on the reconstruction of the discovery of the metabolic pathway for the synthesis of urea [60, 61].

[42] Another example of use of data to generate models and to test them against historical facts was reported by Schummer when analysing the possible causes of the exponential growth of chemical production [109].

Modelling the Evolution of Chemical Knowledge 2

Abstract

We present chemical knowledge as a complex dynamical system emerging from the interaction of the more fundamental material, social and semiotic systems of chemistry. These three latter systems are characterised and further discussed in terms of their constitutive objects and the relations they establish.

In the previous chapter we discussed how chemical knowledge can be described as an emerging property of the historical interaction of the social, semiotic and material systems of chemistry. In the current chapter we provide details on these three systems by discussing the notion of system and by introducing complex dynamical systems as a suitable setting for modelling the evolution of chemical knowledge. We begin by discussing the concept of system.

2.1 Systems and Complex Systems

Although tracing back the origins of the concept of *system* in science constitutes an interesting subject of research, as argued in [110], the idea of system was strongly attached to the mathematical foundations of different sciences, including chemistry, sought for in the Enlightenment. According to Ernst Alfred Cassirer (1874–1945) "for the Enlightenment to 'know' a manifold of experience is to place its component parts in such a relationship to one another that, starting from a given point, we can run through

J. Jost, G. Restrepo, *The Evolution of Chemical Knowledge*, Wissenschaft und Philosophie – Science and Philosophy – Sciences et Philosophie,
https://doi.org/10.1007/978-3-031-10094-9_2

them according to a constant and general rule" [111] (p. 23).[1] Nevertheless, eighteenth-century systems and their mathematical and geometric conception allowed for abstraction, organisation and prediction, which we discuss in Chap. 1 as integral parts of knowledge. John Emmett Lesch (1945–) highlights the use of these systems for instance in botany through the works of Michel Adanson (1727–1806) and Carl von Linné (1707–1778). According to Lesch, these systems not only organised existing knowledge but allowed for indicating knowledge gaps to be filled in. Lesch shows how, for instance, Adanson posed questions such as "Find a known genus of plants (natural or artificial, it does not matter) which occupies an accurate middle point between the family of dogbane and that of borage" [110] (p. 84), which allowed botanists to "be able to estimate how many families or genera were lacking between two distant families or genera whose intermediaries were unknown" [110] (p. 84). In chemistry, therefore, exemplars of systems include affinity tables, the Lavoisian algebraic setting for classes of substances, as well as the Berzelian notation for substances and the molecular structural theory (further details in Sect. 3.2). Affinity tables and the Lavoisian system allowed for estimating reactivities; Berzelian notation for estimating new compounds, as well as the molecular structural theory. Perhaps the most famous chemical system is the periodic system, with its well-known organisational and predictive power.[2] In general, systems become paper tools, using Ursula Klein' (1952–) semiotic metaphor, as they organise knowledge and constitute a mental setting to address past, present and future experiences.

A common feature of systems, recognised by Karl Ludwig von Bertalanffy (1901–1972) is their composition [114]. They are made of objects and their relations. Hence, the Enlightenment idea of a mathematical or geometric structure underlying systems is nicely formalised through a structure defined through its objects and relations. Structures of this sort involve spaces, for instance where objects of a set are endowed with a notion of nearness. In this setting, as discussed in Chap. 1, an Euclidean space is made of elements and a metric relating them. A further example is a graph, for instance a social graph, where social actors are related by friendship, collaboration or other social ties.[3] Bertalanffy identified an interesting feature of some systems, namely their emergent properties resulting from a dynamic interaction of their constitutive parts, which cannot be described in terms of the objects belonging to the system only. He also highlighted the importance of the environment or exterior of the system and especially the interaction system-environment as carrier of information on the functioning of the system [114]. The

---

[1] The seeds of this philosophical stance are evident in Gottfried Wilhelm Leibniz's (1646–1716) conception of a *lingua philosophica* [112, 113].

[2] See [39] for a discussion on the periodic system versus the periodic table. In short, the first is an abstract structure while the second a representation of the first.

[3] In Chap. 1 we also discussed the chemical space, which according to the definition of system can be also regarded as a system. In this sense, the chemical space is also a system making part of the much larger material system of chemistry (see Sect. 2.2.1 for the definition of social system).

study of systems of this sort led to Bertalanffy's general system theory [114] in the 1940s, which set the stage for the incursion of more elaborated mathematical settings of the contemporary complex dynamical systems.

2.2 Chemical Knowledge and its Three Systems

The eighteenth-century mathematician and philosopher Johann Heinrich Lambert (1728–1777) was perhaps the first to conceive a general theory of systems [115], and he distinguished intellectual, intentional, and material systems. That distinction is still meaningful today, and so, for our analysis of the evolution of chemical knowledge, we distinguish between semiotic, social, and material systems. Anticipating more precise definitions below, the social system includes the personal, institutional, and more recently, also artificial actors that drive, enable, influence, or perhaps also impede chemical research. The semiotic system is constituted by the signs expressing, encoding, organising, preserving, and transmitting chemical knowledge, as well as the concepts and theories structuring such a knowledge. The most basic elements of the material system are chemical substances and reactions, but it also includes the devices for creating, manipulating, measuring, identifying and analysing them. Obviously, all three systems not only grow, but also substantially change over time.

It is our basic claim that chemical knowledge emerges from the interaction of these three systems. The thesis that scientific progress results from the interaction of three systems is put forward and developed in [19, 116], where the interacting systems are the social, the material and the epistemic ones. Our setting is different because we consider the semiotic system. What we call the material system of chemistry includes aspects of both epistemic and material systems in the setting reported in [19, 116].[4] Nevertheless, that program is an important inspiration for us.

The evolution of chemical knowledge cannot be properly understood from the isolated analysis of a single one of these systems, nor from their binary interactions. We rather need to combine the analysis of the systems and their interactions with the analysis of their ternary simultaneous relation (Fig. 2.1). The theory of complex systems provides us with an appropriate formal framework for this purpose as it regards the emergence of macroscopic phenomena as the product of multiple relations caused by simple dynamical rules [117]. Importantly, we need to understand the interplay between the structure and

[4] Renn talks about semiotic networks "whose nodes are real objects, material instruments, artifacts, or external representations (such as systems of symbols) and whose edges are the actions, physical transformations, or other relations by which a connection between them is established" [19] (p. 305). As Renn states, they are called semiotic networks "even though they are not limited to signs but rather include the entire material context of action" [19] (p. 305). Our semiotic system, which also entails (as will be discussed) semiotic networks, excludes the material component of Renn's setting.

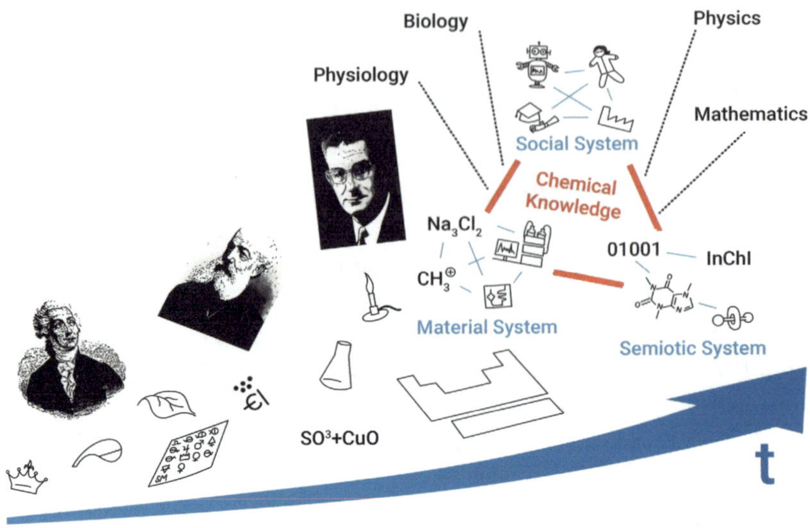

Fig. 2.1 Evolving chemical knowledge as made of the simultaneous interaction of its three systems: social, semiotic and material. Pairwise relations are indicated by continuous lines. Dashed lines represent relations of the systems of chemical knowledge with the external environment. Chemical knowledge is represented as an emergent property of the mutual interaction (bold lines) of the three constitutive systems of chemical knowledge. The evolution of the system is indicated by the temporal dimension. Several historical objects of the constitutive systems of chemical knowledge are also depicted. Representatives of the social system are monarchies (crown), leading chemists such as Lavoisier, Kekulé and Woodward, as well as robots, chemists, academic institutions (academic cap) and the industry. Instances of the semiotic system are affinity tables, Berzelius 1818 doted and barred the chlorine symbol [118] for his Cl^2O^5 as used by Döbereiner in 1829 [119]. SO^3+CuO corresponds to Berzelius 1814 notation for copper(II) sulfate (see Fig. 3.3). The layout of the periodic system, as well as the caffeine molecular structure and a representation of a binary string and InChI are further semiotic objects. Representatives of the material system are a retort, a leaf, an Erlenmeyer flask, a Bunsen burner as well as chemical species such as methenium (CH_3^+) and Na_3Cl_2 (obtained by subjecting NaCl to extreme pressures). A liquid chromatographer and a microfluidic device are also depicted as part of the material system

the dynamics of these systems for a comprehensive picture of the evolution of chemical knowledge as a dynamical complex system.

As a complex system [120], chemical knowledge production tends to be simple while drawing upon more and more complex environments. That is, chemical knowledge maximises its external complexity and minimises its internal complexity. Here, *external complexity* measures the amount of input, information or energy obtained from the environment that the system is capable of handling and processing. For the system, external complexity is input or data complexity. *Internal complexity* measures the complexity of the representation of this input and the schemes handling it by the system. Internal complexity is model or operating complexity. Importantly, what is external and what is internal for the system can change during the evolution of the system. For example, chemists began extracting substances from complex objects of an external world such as minerals, plants and animals. When chemical knowledge evolved, most substances were synthesised internally, that is, providing new substances became technically easier and conceptually better structured. Hence, providing new substances became an internally controlled and thereby considerably simplified process. This in turn enabled a considerable complexity gain of the chemical system.

Complexity measures constitute suitable proxies for analysing the evolution of chemical knowledge. For instance, a crisis of chemical knowledge could be associated with situations where the internal complexity of chemistry is not able to handle its external complexity, forcing a reorganisation and a transition. This leads to ponder, for instance, whether routine research is associated with periods where external complexity is higher than internal complexity. As an example, does the recently found regime of chemical production of new substances between 1860 and 1980 [4] correspond to routine research?[5] In this period the interplay of a powerful semiotics of chemistry based on molecular structures supported by a flourishing chemical industry and by technological advances made it possible to produce, on a regular and systematic basis, new compounds. The semiotic system readily indicated gaps in the chemical space where classes of compounds were to be discovered by already standardised synthetic methods and where advances in other sciences were readily incorporated into the semiotic system. For instance, this period covers the rise of physical chemistry and biochemistry as motivated by input received from biology and physics. This input did not lead chemistry into a crisis, but rather advanced chemical knowledge. This leads to a situation where chemistry was able to handle large amounts of internal and external input and information. That is, the interior and exterior of chemistry provided little "surprise" for that chemistry. Such a welcoming of information was based on a "well-prepared" chemistry whose knowledge was able to handle the input with little perturbation of its structure. What happened in terms of complexity in the third regime?[6] Did the rise of organo-metallic chemistry increase the internal complexity of chemistry? If so, was this driven by the internal structure of chemistry or did rather something happen at the exterior of chemistry, perhaps in physics and engineering? To which extent was the bio-organic part of the third regime, between 1990 and 2015, again "welcomed" by the existing chemical knowledge? Solving these questions requires further research. Nevertheless the data and methods are available to start such an investigation. This we show in several of the following chapters, especially in Chaps. 5 and 6.

[5] In Nicholas Rescher's (1928–) studies on scientific progress [121], scientific growth with annual production rates (r) are identified as first rate topics, when the growth is linear, and as very important ($r \sim 1.25\%$), important ($r \sim 2.5\%$), significant ($r \sim 3.75\%$) and routine research topics ($r > 5\%$), when the growth is exponential. In [4], we found that the proto-organic regime (1800–1860) was close to "significant," while the organic regime (1860–1980) was roughly "routine research." A closer look at the organo-metallic regime shows that its first decade (1980–1990) was "very important" and that subsequent years (1990–2015) have been close to "routine research" (Fig. 1.3).

[6] See previous note.

In the following subsections we describe the constitutive systems of chemical knowledge, how they are compounded and the relations holding them together.

2.2.1 The Social System

Chemists interact, cooperate and compete, and this is organised and canalised by formal or informal institutions. The objects of the social system of chemical knowledge involve actors such as people, academic and scientific societies, committees, enterprises, industries and other forms of social organisation. More recently, also computational objects such as robots and artificial intelligence technologies become actors, and we should include them also in the social system of chemistry. The social role of artificial intelligence technologies was recently discussed, for instance, at the US Patent and Trademark Office as a consequence of patent applications for devices created by software [122]. Although the US Patent Office ruled that artificial intelligence cannot be a legal inventor, the fact of having discussed the issue shows the social status this technology has gained. Moreover, nothing restricts further discussions and possible changes of that decision. Another example of the social role of artificial intelligence is the publishing of chemical literature without human edition and only based on data-driven analyses of reported literature [123]. These examples complement the ongoing robotisation of chemical synthesis [124–127].

The objects of this system, held together by economic, political, cultural, academic and other relations constitute the social system of chemistry.

This system is the basis for the social perception of chemistry, at a large social scale, and to the self image of chemists [128]. The social system relates to ethical issues [129],[7] such as the moral of chemical synthesis, chemical weapons research, environmental pollution, chemical accidents, unintended bad "side-effects," moral issues related to the improvement of material conditions of life by chemical means, and freedom of research. Moreover, one can pose the question whether chemistry should advance for the sake of humans or for the very sake of chemical knowledge [130].[8] Another social aspect related to the evolution of

[7] In this respect, Schummer has set up the basis for moral judgements of synthetic chemistry in a public discourse and the framework for chemists to reflect upon the moral relevance of chemical activities [130].

[8] Interestingly, basic notions of ethics, such as responsibility; can be modelled as directed hypergraphs (Chap. 6). In general, x is responsible for y to z, where x is a subject or agent of responsibility, y are the consequences of x's actions (or omissions) and z is the institution to which x feels or are made obliged to justify its actions related to y [130]. Then, responsibility is modelled as two binary directed relations: $x \rightarrow y$ and $x \rightarrow z$. As x, y and z may be subsets, the most general framework for responsibility is a model of two directed hypergraphs, $\{x\} \rightarrow \{y\}$ and $\{x\} \rightarrow \{z\}$. Two examples of responsibility are: Carl Djerassi (1923–2015) is responsible for contraception to humanity. Or the United States Chemical Warfare Service (as an institution made of a set of people) is responsible for Napalm to humanity [131]. A further example showing the different levels of chemical responsibility is that of a synthetic chemist being responsible for her/his synthetic product or for her/his chemical

chemical knowledge is the collective action problem; that is, how social actors, who could cooperate for the sake of chemical knowledge, decide not to do it because of conflicting interests, for example scientific rivalry or industrial secrecy.[9]

Even if we describe several social objects and their relations, we do not ignore their historical and epistemic connections. For instance, if we talk about chemists as part of the social system, by "chemists" we mean scientists acting upon matter , industrial advisers and entrepreneurs, as well as descendants of apothecaries, pharmacists and metallurgists. We talk about the evolution of their social perception and of their interaction with social structures such as nations, political institutions and their ruling bodies, academia, industry, educational systems, public media, among others.

2.2.2 The Semiotic System

Chemistry is a system for the generation, preservation, organisation and transmission of scientific knowledge. From a functional perspective, chemists, like other scientists, perceive objects and processes through or as signs, and form signs to interpret and communicate their findings, and to express the systematic relations they discover and to form categories and norms. In particular, signs constitute both the means and the units of communication. From an operational perspective, analogies and similarities, classifications and distinctions all are results of operations with signs. These functional and operational sign processes are the subject of semiotics. The generation, transmission and understanding of meaning leads us into the realms of semantics and pragmatics.

In the semiotic theory created by Charles Sanders Peirce (1839–1914), signs are linked to objects and interpretants [133, 134]. For instance, a colourless liquid wetting material, quenching our thirst, flowing in rivers, boiling at 100 C, dissolving table salt, plus several other features, is the object we (interpretants or sign users) attach to the sign "water" in English. This sign is not randomly selected, as interpretants long ago considered that the wetting capacity of this liquid was relevant to the intended signification.[10]

Thus, when we analyse the semiotic system of chemistry, we should look for Peirce's objects, signs and interpretants and their relations. That is, we have to link the signs, which constitute the semiotic system, with the objects in the material and the interpretants in the social system.

---

knowledge to the synthetic chemistry community, to the chemistry community at large and to the society.

[9] An issue related to current cooperation is that of bias when declaring contributions to scientific work [132].

[10] *Water* contains the proto-Indo-European root $*$*wed* or $*$*wod*, making it a cognate with *wet*, and which is also ethymologically related with the German *Wasser* [135]. In contrast to English, in Spanish, for instance, the feature to highlight of this liquid is its connection with rivers, as evidenced in the Indo-European root $*$*akwa* [136], which via the Latin *aqua* leads to *agua* in Spanish.

The semiotic objects of chemistry include substances and reactions. Interpretants (chemists) have come up with different signs for the former, for example descriptions of their extractions and origin [29,71]; empirical formulae to signify composition; secondary qualities [26, 27], chemical, physical and biological properties,[11] to signify a space of properties and substance identities, to which also composition is added; substance names to identify and highlight discoverers (Grignard reagent) or geographic spots (Prussian blue); alchemical symbols to signify mystical connections or to encode important properties; radicals signifying invariant matter pieces under chemical reactions [29]; functional groups to indicate types of chemical reactivity; atoms, initially invariant signs as radicals and latter signifying material particles; spectra as signs for identity and several others. Some of the signs for reactions have been affinity tables signifying the possibility and readiness of reactions; reaction classes to differentiate types of transformations; reaction mechanisms to explain reactions; and potential energy surfaces to forecast reactions as well as to understand their energetics.

From the interplay between reactions and substances new signs result, for instance those for chemical bonds, as well as periodic systems as signs of abstract relations among substances. Further high order relations between objects require new signs enriching the semiotic system of chemistry. Although the interpretant is transversal to the interaction with objects and signs, we highlight here that her/his interaction is often mediated by instruments and apparatus, along with her/his theories allowing for the desired signifying act of the semiotic process.[12] This latter relation is pointed out by Stephen J. Weininger through the formula H_2O, which may represent a substance, a molecule, clusters of them, a condensed phase, an acid or a base, a nucleophile or an electrophile and this is only determined by the accompanying descriptive text as provider of context [138].

A further aspect of the semiotic system of chemistry entails the narrative used by chemists over the evolution of their discipline. Here, by narratives, we mean, following [87], "higher-order structures of how stories are related to each other. They constitute patterns of story-telling in both their temporal and structural dimensions." To which extent are chemical narratives similar to those of other sciences or the humanities? How does the incorporation of new semiotic objects or semiotic relations affect chemical narratives?

As well as in the social system, the objects and relations constituting the semiotic system are traversed by their historical and epistemic ties. For instance, "oxygen," as

[11] Locke' secondary qualities [27] and properties, for example of substances, come from observing the interaction of these materials with their environments. Thus, chemical properties result from the mutual interaction of substances; physical properties, from mechanical forces and electromagnetic fields acting upon substances; and biological properties from the interaction of substances with living systems [26].

[12] Contexts are also relevant in the semiotic system as they constitute sign ensembles allowing for meaning generation. Interesting discussions on the semiotics of chemistry are found in [31, 137].

conceived by Lavoisier must be regarded as the sign encoding the nature of acidity and what is given off when compounds are formed by combustion [87].[13] Today, even if the sign is the same as the Lavoisian one, we must be aware that the sign is endowed with novel features, ranging from its quantum structure up to its relation with the ozone layer and our presence on Earth.[14]

2.2.3 The Material System

The basis of chemical knowledge are substances and reactions (Fig. 1.2). As chemists witness, devise, control and theorise upon substances and their transformations; semiotic and material systems are strongly related to each other. For example, properties are semiotic objects as they are observed and used as communication tokens by chemists. Properties also emerge in the material system, as they result, for example from the relation between substances, measurement devices and theories underlying measurements. The semiotic objects of substances, reactions and their classifications can only result from the experimentation upon substances. Atoms, bonds and molecules are examples of semiotic objects resulting from chemical experiments, which were subsequently, through more experimental evidence, introduced as material objects and have motivated the rise of novel semiotic objects [71, 91], for instance reaction mechanisms.

Hence, the objects of the material system are substances, reactions, atoms, bonds, molecular species, materials (including organised materials such as plants and animals), reaction conditions and the technologies used to interact with those objects, from chemistry glassware up to automatised spectroscopic devices and robots for chemical synthesis

[13] Interestingly, Hasok Chang (1967–) has made the point that epistemic objects are not only related by historical paths, but often rather by different emphasis on some particular features shared by objects in their evolution [139].

[14] A similar enrichment of a chemical concept occurred in 1754, when Guillaume-François Rouelle (1703–1770) introduced the concept of "base" and widened that of salts. Before Rouelle, salts have meant neutral and soluble compounds. By using the concept of bases, he called salts the product of the reaction of acids and bases. In so doing, salts gathered acid, basic and neutral compounds, which could be insoluble in water [35].

and analysis.[15] Objects of the material system are held together by chemical reactions,[16] measurements upon substances and associations between substances or molecular species and materials where they are extracted from.

As well as we discussed in the social and semiotic systems of chemistry, objects and relations of the material system are not to be regarded as static entities appearing in the history of chemical knowledge. They are rather tied by evolutionary paths and contingencies, which provide contexts for their historical understanding. This is evident, for instance, when referring to "reactions." Before mid eighteenth century it was thought that reactions only occurred among liquids or solids. Gases, as championed by Jan Baptist Van Helmont (1577–1644) could not take part in reactions [35]. However, in 1756 Joseph Black (1728–1799) showed experimentally several reactions with gases, widening the reaction concept [35]. In a late nineteenth- or early twentieth-century chemistry setting, besides considering Black's ideas on substances in different phases able to react, a chemical reaction entails a chemical transformation under controlled conditions of temperature, pressure, pH and stirring. These conditions are to be regarded as continuous and homogeneous variables, which act in a vessel, whose dimensions are irrelevant as well as the material from which it is made. In contrast, although "reaction" refers to the same process in a late twentieth- and twenty-first-century framework, now if the amount of reacting substances is very small and the vessel is also small, say at the nanolevel, then substance and vessel shapes are key for understanding the process. Moreover, the material from which the vessel is made of becomes an important part of the "reaction." Likewise, the nanoscale requires abandoning the idea of homogeneous reaction conditions.[17]

It may also happen that the same substance or material adopts different names (or signs in semiotic terms) across history. For example, alcohol (actually a dilute ethanol solution in water) took the names *aqua ardens*, burning water[18]; *aqua vitae*, water of life, in the 12th century [35, 142] and finally its current name after the sixteenth century, when

[15] When describing the social system of chemical knowledge we have discussed how robots and artificial intelligence technologies are also part of that system. Incidentally, we see here a case where something that traditionally was thought of as part of the material system, mechanical and computational devices, became so advanced that it should now be reconsidered as belonging to the social system. This is a case of a transition between two systems, and it shows that the boundaries between them may be somewhat fluid or permeable. In the other direction, the role of animals in research has also been and is still contested, and in the twentieth century, we have even seen terrible examples where human beings became objects or targets of chemical experiments.

[16] Networks of chemical reactions giving place to new functions such as life, are also relations considered by the material system; which are currently studied under the name of systems chemistry [140].

[17] Further information on reactions at nanoscale levels as well as their theoretical settings are found in [141].

[18] As it burns at low temperatures avoiding to inflame the material on which it is placed. Incidentally a solution of this sort still preserves the meaning of *aqua ardens* in Spanish, where it is called *aguardiente*, which results from connecting *agua* and *ardiente*.

Philippus Theophrastus Bombastus von Hohenheim (1493–1541), who called himself Paracelsus, derived its name from the Arabic *al kohl*. This term had been originally applied to powdered antimony sulfide used for darkening the eyebrows [35]. By the sixteenth century the term came to mean any fine powder and it was then used to signify the essence of things. In this case it was the essence of wine, therefore Paracelsus referred to *alcool vini* or *alcohol vini* [35]. Later the word vini was dropped and today alcohol means not only ethanol but the whole class of compounds to which ethanol belongs.

Evolution of the Constitutive Systems of Chemical Knowledge

3

Abstract

The evolution of the three constitutive systems of chemical knowledge is presented. In the social system, we see the growth of and the changes and developments in the chemical community and its institutions and the network-like framework connecting the various actors. The evolution of the semiotic system is presented as a shift of mathematical frameworks ranging from ordered sets to algebraic, topological and geometric settings. We show how the evolution of the material system corresponds to an expanding hypernetwork of chemical reactions that is related to the technologies and laboratories allowing for its expansion.

In the previous chapter we characterised the three constitutive systems of chemical knowledge. Here we discuss their historical dimension by analysing the dynamics of each system from a formal and computational perspective. The purpose of this approach is to set the stage and to motivate the computational and formal methods that we discuss in the second part of this book.

3.1 Evolution of the Social System

Scientific paradigms spread, become dominant and eventually die out in social systems [48, 143]. The perspective developed here, however, suggests a more refined picture than that of *thought collectives* [48] or *paradigm shifts* [143]. In fact, we look at the interactions between three different systems, the social, the semiotic and the material one, with their

J. Jost, G. Restrepo, *The Evolution of Chemical Knowledge*, Wissenschaft und Philosophie – Science and Philosophy – Sciences et Philosophie,
https://doi.org/10.1007/978-3-031-10094-9_3

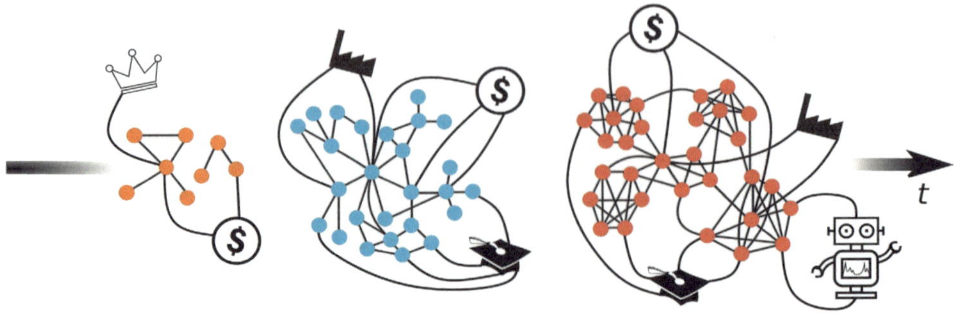

Fig. 3.1 Evolution of the social system of chemical knowledge. These networks represent relationships among chemists (dots), where other social actors such as monarchies (symbolised with a crown), economic, industrial and academic factors come into play. The role of robots in current chemistry is highlighted in the network on the right. Details on the temporal relationships among social actors are provided in Sect. 4.6 and in Fig. 4.2

own intrinsic dynamics. Social processes, semiotic conventions, and chemical discoveries drive each other. In this section, we shall look at the social system.

The social system evolves not only because people play their social role but because some of them, such as Lavoisier, Liebig, Kekulé or Woodward, have created new schools of thought, which, besides constituting a dynamical social network (Fig. 3.1), are associated to semiotic and material networks, where concepts, theories, technologies and innovations relate with each other (Fig. 2.1).[1] Schools of thought may dominate the scientific thinking for a while, both because of their intrinsic insights and because the leaders have obtained influential positions in the social system. These influences are therefore related to scientific luminaries' lifespan.[2] That is "A new scientific truth does not triumph by convincing its opponents and making them see the light, but rather because its opponents eventually die, and a new generation grows up that is familiar with it" [145] (pp. 33–34). It has been found that once luminaries pass away, their scientific fields surge by the multidisciplinary ideas of newcomers [146]. Even though single prominent scientists do not constitute collectives of their own, they may lead them, because their scientific success assures them relevant positions in the social network with many followers. In this way, luminaries may become gatekeepers regulating access to others and driving the evolution of their fields [146].

A fundamental question on the evolution of the social system of chemistry is the growth of the community (represented as an expanding network in Fig. 3.1), which has been a

[1] The relationships, growths and couplings of social, semiotic and material networks constitute an interesting case to be modelled with multilayer networks (see Chap. 6).

[2] Once Max Karl Ernst Ludwig Planck (1858–1947) stated that "Eine neue wissenschaftliche Wahrheit pflegt sich nicht in der Weise durchzusetzen, daß ihre Gegner überzeugt werden und sich als belehrt erklären, sondern vielmehr dadurch, daß die Gegner allmählich aussterben und daß die heranwachsende Generation von vornherein mit der Wahrheit vertraut gemacht ist" (p. 64) [144].

question addressed by historians through the analysis of membership records to chemical societies and to the industry [147], as well as by analysing the launching of new journals, the publication of dissertations and the creation of new societies [5, 148, 149]. These figures show that by 1900 there was an unusual surge in membership to European chemical societies, most of them launched around the turn of the twentieth century [150]. This was particularly evident for the French and German societies [151, 152], which was presumably caused by the enormous popularity of chemistry in the second half of the nineteenth century,[3] which we discuss in this book as motivated not only by social reasons, but by semiotic and material ones. A further approach to quantifying the size of the chemical community is through bibliometric tools, where the number of chemists is gauged through their publications. Schummer has followed this approach [109].[4] His results cover the period 1900–1995 and show an exponential growth,[5] whose interplay with the exponential growth of the chemical space has to be analysed.[6]

Other social structures contribute to the evolution of the social system, for example new organisational structures associated with chemical practice have shown up, ranging from isolated alchemists; seventeenth-century organised groups, academies and societies[7] to nineteenth-century institutions, which combined teaching and research in more collaborative laboratory practices (Fig. 3.1).[8] Further organisational structures involve interactions of chemists with academic and industrial players, as well as with artificial intelligence technologies. This social evolution has also been attached to national and international competition and has responded to social demands such as procuring work protection or

[3] According to Helge Stjernholm Kragh (1944–), the success of chemistry in the nineteenth century arose from the peculiar situation chemistry held among the sciences, as it found use in a large number of diverse fields [153] (p. 329).

[4] From a historical perspective, not only chemists are to be included but apothecaries, miners, metallurgists and amateurs, among others [154]; which cover the period before the establishment of chemistry as an academic discipline in the nineteenth century. Determining the size of this community poses challenges, which may be overcome by using scientometric approaches, as well as several of the tools discussed in Chap. 6. Of course, we need to be aware of the fact that such computational approaches depend on implicit assumptions about who counts as a chemist. For scientometric data, this is obvious, as there a chemist is a coauthor of a paper in one of the chemical journals included in the database, or perhaps of a patent application.

[5] Schummer also analysed the interesting ratio of the number of new substances per chemist and found that in the 1950s and 1960s there was a surge of this ratio. This is presumably related to the adoption of spectroscopic methods for molecular structural determinations in the 1950s. In this respect see references [155, 156].

[6] Early studies of this interplay are addressed by Schummer in [109].

[7] Some of these collectives are the private academies of Mersenne or Justel in Paris, the Invisible College to which Boyle belonged and that met in Oxford and London and formal groups such as the Royal Society in London (1662) and the *Académie des Sciences* in Paris (1666) [35].

[8] An example of these teaching and research institutions is Justus Freiherr von Liebig's (1803–1873) model for the nineteenth-century laboratory, which was taken as a worldwide referent [29] (p. 203).

developing chemical products to improve national economies or to assist their military purposes [137]. The ongoing computerisation of chemical activities relates to new social structures for chemistry, which involve not only more interdisciplinary teams but the appearance of non-human actors in the system, as discussed before (Fig. 3.1).

3.2 Evolution of the Semiotic System

Over the course of history, chemistry has developed several intricate symbolisms and concepts, each adapted to the chemical knowledge of its time, encoding that knowledge, and thereby profoundly shaping chemical thinking. Although since the times of Lavoisier, chemistry no longer has gone through such dramatic periods as other sciences like physics or biology, its semiotic system witnessed perhaps more profound changes than those of most other disciplines. At the same time, it participated in general trends of the semiotic system of science as a whole.[9]

Thus, besides the semiotic evolution of chemistry evidenced in the language used in each period, ranging from Egyptian, Mesopotamian languages, Greek, Arabic, to Latin, French, German and currently English, there are subtle issues pertaining to the semiotic system of chemistry.

The central question has always been how to have signifiers (signs) that express the essence of the signified and relate them to their referents (objects), which, coupled to the social dynamics of the advance of chemistry, become the causes of the evolution of the semiotic system of chemistry. For instance, sixteenth-century quantitative methods led to the introduction of yields as a concept conveying relevant information of chemical reactions [35], although general conservation laws that could systematically guide chemical thinking were only formulated much later. In particular, conservation of mass was first formulated and applied as a guiding principle by Lavoisier in the late eighteenth century, see the discussion of his role in Chap. 4. Another example of the semiotic evolution of chemistry includes eighteenth-century affinity tables[10] (Fig. 3.2), which

[9] With this we do not mean that chemistry has not had post-Lavoisian watershed events. Other examples include the development of atomic theory leading to the universal adoption of structural chemistry thinking [91], a movement that started at the dawn of the nineteenth century and that still shapes the chemical way of thinking and of addressing the material world [26, 31, 137].

[10] An instance of these tables is shown in Fig. 3.2, which is to be interpreted column by column in such a way that substances below the top one are ordered according to their strength of affinity regarding the top substance. As an example, the first column (left hand side), indicates that acid spirits (top substance) have strongest affinity for fixed alkali salt, then for volatile alkali salt, absorbent earth and finally for metallic substances. This indicates that fixed alkali salt reacts more favourably with acid spirits than the other substances down the column and that it displaces all the substances below it from their existing combination with acid spirits. Likewise, this column informs that volatile alkali salt displaces absorbent earth and metallic substances from their combinations

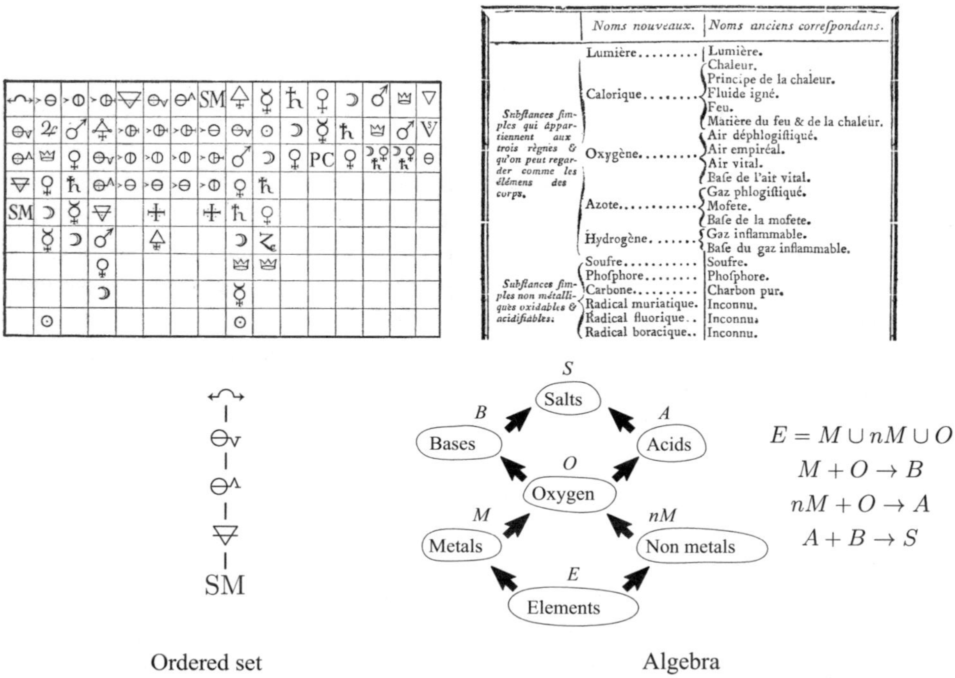

Fig. 3.2 Evolution of the semiotic system of chemical knowledge. On the left, Geoffroy's *Table des differents rapports observés entre differentes substances* (1718), which corresponds to ordered sets of substances (below left). Reproduced from Figs. 6 and 7 of [112]. On the right, excerpt from Lavoisier's *Traité élémentaire de chimie* (1789, p. 192) [165] and the algebra of sets in which his system can be framed (below right) [112]

depicted information about reactions[11] and gave place to systems emphasising substances and currently molecules. Affinity tables are collections of ordered sets of reactions with substances identified by alchemical symbols [158], which moved to an algebra of classes of substances and to a new system of nomenclature introduced by Lavoisier (Fig. 3.2) [112]. The dawn of the nineteenth century turned substance constitution central for the semiotic system, with elements represented by Roman alphabetical characters[12] and

with acid spirits but does not displace fixed alkali from its combination with acid spirits. More details on the interpretation of this table in [157, 158].

[11] An interesting approach to chemical formulae as bearers of relational information in chemical reactions is that of Benjamin Collins Brodie (1817–1880) [159, 160], who claimed that the chemical question is not what water is but how it behaves during interaction with other substances [161]. Similar transitions from an essentialist to an operationalist perspective can also be seen in other scientific disciplines and are amenable of being formalised in terms of category theory [162].

[12] On an alternative system of symbols for chemical elements, see for instance the Greek symbols used by Brodie in his calculus of chemical operations [159, 160].

compound constitution by an algebraic system of elements, a turn championed by Jöns Jakob Berzelius (1779–1848) [71] (Fig. 3.3). By mid-nineteenth century the system moved to a topological system of atoms and bonds, that is of molecular structures (Fig. 3.3), which is currently combined with a geometric system associated to crystallography and quantum chemical results, as well as descriptions of nanostructures (Fig. 3.4). The semiotic of substances has included iconic signs, such as those used in alchemical texts and pre-Lavoisian documents (Fig. 3.2), as well as molecular structures [31, 137, 138]. It also includes symbolic linguistic signs, such as substance names. Interestingly, the computational era directs the semiotic system back to text signs as evident in the use of SMILES and InChIs to identify compounds (Fig. 3.4).[13]

Chemists, perhaps motivated by their desire of unveiling new substances, traditionally report positive results, that is reactions leading to target products.[14] However, behind those successful stories, lots of failed reactions lie on lab notebooks. Chemists experiment much more than what they report and the community, after decades of concern [172–174], has begun to handle negative experiments or dark reactions. This awareness has led to the introduction of failed reactions, and its associated metadata such as reaction conditions, into the semiotic system. Evidence of this move is the open-notebook science initiative [174], where experimental results are posted, commented and discussed in blogs and wikis. Also leading journals [175] are moving in the direction of reporting failed reactions along with successful ones. The hand-in-hand use of failed and successful reactions is not only related to completeness of the super structure of reactions associated to the chemical space,[15] it has also proven instrumental for the production of particular chemical targets [176].

In a linguistic framework, the evolution of the semiotic system can be regarded as the evolution of a language [177], that is based on grammatical rules. These rules change

[13] InChIs are explained in footnote 33 (Chap. 1). SMILES (Simplified Molecular-Input Line-Entry System) are string representations of molecular structures. Examples of InChI and SMILES are shown in Fig. 3.4.

The advent of machine learning has made it possible to think of treating molecular structures and human molecular drawings directly as computational objects [163, 164]. Efficient algorithms for these purposes would overcome the problem of representing molecular structures as strings of characters required for their computational storage and processing for the computational infrastructure of the late twentieth century.

[14] A further reason is funding, as it becomes more convincing to show positive results than failed attempts. This is evident, for instance in Chinese alchemy, were public attempts by 144 B.C. to make gold for the emperor failed and contributed to the bad reputation of alchemy [35].

[15] By "super structure of reactions" we mean the wiring of the chemical space through chemical reactions, which formally can be either successful, that is linking some substances (substrates) with others (products); or unsuccessful (failed or dark reactions), proving that under the conditions of experimentation it is not possible to link the substrates in question with any other substance of the chemical space or at least with those expected according to the chemical semiotics by the time of experimentation.

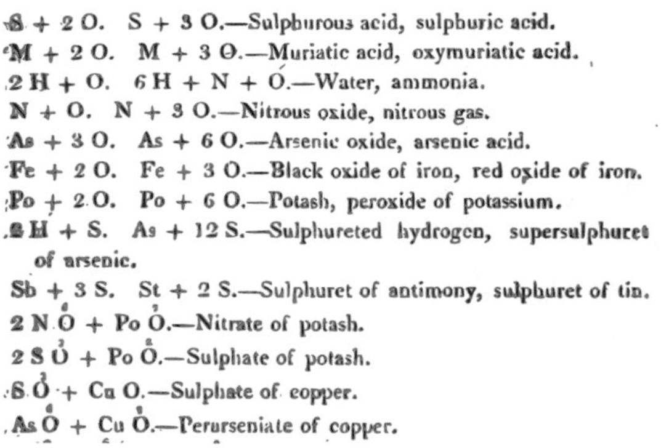

S + 2 O. S + 3 O.—Sulphurous acid, sulphuric acid.
M + 2 O. M + 3 O.—Muriatic acid, oxymuriatic acid.
2 H + O. 6 H + N + O.—Water, ammonia.
N + O. N + 3 O.—Nitrous oxide, nitrous gas.
As + 3 O. As + 6 O.—Arsenic oxide, arsenic acid.
Fe + 2 O. Fe + 3 O.—Black oxide of iron, red oxide of iron.
Po + 2 O. Po + 6 O.—Potash, peroxide of potassium.
2 H + S. As + 12 S.—Sulphureted hydrogen, supersulphuret of arsenic.
Sb + 3 S. St + 2 S.—Sulphuret of antimony, sulphuret of tin.
2 N O + Po O.—Nitrate of potash.
2 S O + Po O.—Sulphate of potash.
S O + Cu O.—Sulphate of copper.
As O + Cu O.—Perarseniate of copper.

$$\begin{aligned} \textit{Oxidum cuprosum} &= \mathrm{Cu + O} \\ \textit{Oxidum cupricum} &= \mathrm{Cu + 2O} \\ \textit{Sulphuric acid} &= \mathrm{S + 3O} \\ \textit{Sulphate of copper} &= \mathrm{CuO} + \mathrm{S}\overset{3}{\mathrm{O}} \\ \textit{Persulphate of copper} &= \mathrm{Cu}\overset{2}{\mathrm{O}} + 2\mathrm{S}\overset{3}{\mathrm{O}} \\ \textit{Alum} &= 3(\mathrm{Al}\overset{2}{\mathrm{O}} + 2\mathrm{S}\overset{3}{\mathrm{O}}) + (\overset{2}{\mathrm{Po}} + 2\mathrm{S}\overset{3}{\mathrm{O}}) \end{aligned}$$

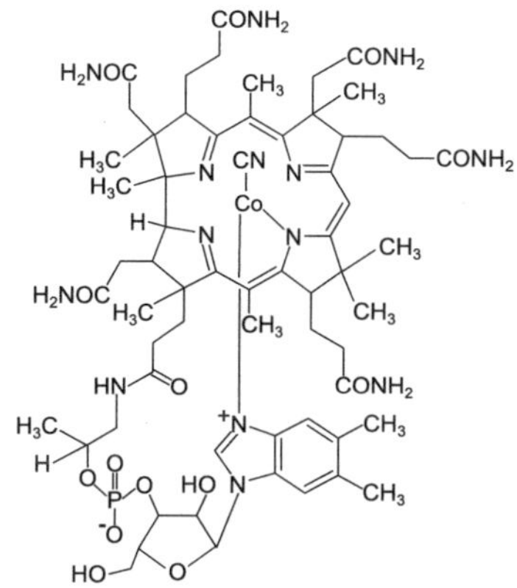

$$(E, \tau)$$

$$E = \text{Set of elements}$$

$$\tau = \text{Family of subsets of } E$$

Algebra Topology

Fig. 3.3 Evolution of the semiotic system of chemical knowledge. On the left, excerpt from 1814 Berzelius' algebraic nomenclature system [166] (p. 363). Below the algebraic equations for some substances, as expressed by Berzelius in 1814 [166] (p. 52). On the right, molecular structure of vitamin B12 (adapted from [167]) and below the topological setting for molecular structures, which can also be expressed in graph theoretical terms, where a molecular structure corresponds to (E, b), with E being the set of elements and b the set of bonds [168]

based upon cultural evolution such that they compete with each other making that new rules rise while old ones fade away [178]. For instance, in the Lavoisian-Berzelian system, grammars ruled over formulae producing new formulae of the language. Likewise, the current language of chemistry is mainly based on grammars operating over molecular structures (see Chap. 6).

Although the chemical semiotic system has strongly evolved in the direction of an ontology of substances and molecules,[16] the historical ontological question of chemistry still remains: is chemistry about substances or about reactions? [179]. Historical supporters of the ontology of reactions are Geoffroy [138], Brodie [180], and more recently Earley [181–184], van Brakel [185], and Stein [186], with Schummer providing an interesting

[16] Weininger claims that the rise of organic chemistry in the nineteenth century was marked by a semiotic preoccupation with signs themselves, weakening the reciprocity between sign and experiment [138].

InChI=1S/C2H6O/c1-2-3/h3H,2H2,1H3

OCCc1c(C)[n+](cs1)Cc2cnc(C)nc2N

$$(E, d)$$
$$E = \text{Set of elements}$$
$$d = \text{Metric on } E$$

Geometry

Fig. 3.4 Evolution of the semiotic system of chemical knowledge. Left and centre plots correspond to geometric characterisations of chemical species. On the left, the ORTEP (Oak Ridge Thermal Ellipsoid Plot) diagram (X-ray crystal structure) of a Copper(II) N-heterocyclic carbene complex, adapted with permission from Fig. 2 in [169]. Copyright 2021 American Chemical Society. The image at the centre corresponds to the *in situ* atomic-resolution detection of a forming NaCl nanocrystal in a carbon nanotube, where chlorine and sodium atoms form a grid. Figure adapted with permission from graphical abstract in [170]. The two textual representations on the right, correspond to the InChI string [76] for ethanol (top) and to the SMILES [171] string for thiamine (bottom)

ontology for chemistry based on substances and reactions in a network of dynamical relations (Fig. 1.2), where substances and reactions mutually define each other at the experimental and theoretical level [179]. This is the approach we take here, which can be formalised through category theory,[17] where objects are defined by their relations [162].

Remarkably, however, the essence of the Berzelian system that represents substances by their elemental composition is still used and useful, in spite of its shortcomings just described, like not representing the topological molecular configuration and therefore not being able to distinguish between isomers, for instance. This sort of semiotic stability distinguishes chemistry from other sciences, as we will discuss in the following lines. Nevertheless, in establishing a systematic and long lasting nomenclature, chemistry was preceded by biology, with the so-called natural system of Linné for the classification of biological species. It is still used for biological taxonomy even though its conception was pre-Darwinian.[18] But taxonomy nowadays is only a small subdiscipline of biology. In this

[17] For an introduction, see for instance [187].

[18] In fact, the concept of a species has profoundly changed since then. Linné conceived a species as a lowest unit in an invariant hierarchical system that can be characterised in terms of some specific morphological difference to other such units. From a Darwinian perspective, species are

comparison of biological and chemical nomenclature one is tempted to analyse the case of molecular biology. In this biological subdiscipline nomenclature depends on chemical classification systems, although the units are not atoms, but rather molecular units, like nucleotides or amino acids. But this again leads to problems, for instance for the gene concept which should intertwine coding with regulation and function to be of biological significance [192]. Also, a string of, say, 300 amino acids does not typically yield a protein, because it needs to be able to fold properly to become functional. One therefore considers basic folding motives like α-helices or β-sheets [193]. The organisation of data about genes, RNA molecules or polypeptides in biological data bases requires still other principles of biological nomenclature.

Thus, chemistry is in a better position than biology, because, notwithstanding its shortcomings, describing a substance through its atomic composition still provides a universal scheme that can then be refined by including topological motives, geometric conditions, or coarsened by symbols for particular combinations or patterns, as described above.[19] The contrast to physics or mathematics is even starker. Chemistry is essentially a *material* science,[20] not with the meaning of the technical field of modern material science that is concerned with the construction of nano- or micro-materials with specific properties, but simply because chemistry is about objects and processes with a specific physical identity.[21] Physics, in contrast, is to a large extent a *structural* science, and mathematics even is exclusively so. That means that they are concerned with formal structures or relations that can be materialised very differently.[22] For instance, there is

the leaves or endpoints of some phylogenetic tree where the hierarchical organisation is replaced by common ancestry. Willi Hennig's (1913–1976) cladistics [188] derives this organisation from a logical analysis of shared characters. While there is considerable debate among biologists about the appropriate definition of a species (see for instance [189]), essentially because all theoretical notions have to admit many exceptions in biological reality, most contemporary species concepts can be seen as variants of the population based concept of Mayr (1904–2005) [190, 191] which considers a species as a population capable of interbreeding and reproductively separated from other species. In particular, this is important for understanding speciation processes. While Linné's scheme does not and cannot capture this dynamical aspect, it is still universally used by taxonomists.

[19] This chemical semiotic stability is nonetheless imperfect, as we have discussed in Chap. 1, where we mentioned how InChIs, which encode Berzelian composition, fail to represent inorganic substances. Likewise, the different systems of nomenclature for different classes of compounds and for several subdisciplines of chemistry, devised through the history of chemistry, evidence the constant change of the semiotic system of chemistry.

[20] An interesting account of this perspective is found in [26].

[21] As discussed in [26], chemical species (substances, molecules, clusters, nanoparticles and more) are inherently defined by their relations with others, which define the chemical context, or chemical universe, in which all narratives of chemistry live.

[22] Schummer [26] argues that physics is also a material science, but unlike chemistry, the subject of study is focused on how matter interacts with mechanical forces, thermodynamic conditions and electromagnetic fields. In chemistry, the focus is on how matter interacts with other matter realisations, that is how chemical species relate one with each other [26].

the field of econophysics which analyses economic data with statistical tools, from time series analysis or other disciplines. These tools can be, and have been used, to also study data sets from very different fields. But there is no corresponding field of econochemistry, because economic systems are based upon human actors and institutions, rather than upon particles or molecules. In particular, neither physics nor mathematics has achieved or can conceivably achieve an essential semiotic stability, as that we have here described for chemistry, that can cover the entire discipline. The possible structures and relations are simply too diverse for that purpose.

Here we have presented a small set of cases evidencing the evolution of the semiotics of chemistry. It is not intended to be complete, it rather may be taken as a motivation for more specialised and detailed studies on the semiotics of chemistry. Nevertheless, in Sect. 4.6 we provide further instances of the semiotic evolution that include mergings, splittings and refinements.

3.3 Evolution of the Material System

A recently found evidence of evolution of the material system is the detection of three clearly demarcated statistical regimes in the production of new chemicals over the history (Fig. 1.3) [4].[23] The proto-organic regime (1800–1810) corresponds to the most unstable period in the annual production of chemicals and is characterised by a relatively high proportion of inorganic chemicals, when contrasted with organic compounds. The sharp transition to the organic regime occurred around 1860 with a drastic stabilisation of the annual production of new substances, which lasted until 1980, when a second drastic stabilisation took place. The organic regime is characterised by a strong emphasis on organic chemicals, while the transition around 1980 by an emphasis on organo-metallic substances. Besides the expansion of the chemical space, other evidences of the evolution of the material system come from changes in chemical instrumentation and the appearance of new ontological species. For instance, the development and adoption of spectroscopic methods in the 1950s [155, 194] coincided with a second rapid report of new substances (Fig. 1.3).[24] By analysing Fig. 1.3 it is observed that the surge of new compounds went accompanied by a small drop in the proportion of synthesised substances. Therefore, indicating a small increase of extracted substances after the 1950s. The question that arises is whether this was motivated by the adoption of spectroscopic methods.[25]

[23] See footnote 5 in Chap. 2.

[24] The first rapid discovery occurred between 1870 and 1910 (Fig. 1.3) and is accompanied by the adoption of the structural theory as a paper tool [71] or blue-print ideology.

[25] Before the advent of spectroscopic techniques, a new substance, either synthesised or extracted, was subjected to several chemical reactions as well as their products in order to determine the molecular structure of the new substance. This implies that every new substance led to several other new ones. Spectroscopic methods bypassed those reactions leaving each new substance with

New ontological species have appeared, such as reaction intermediates,[26] interlocked molecules and nanostructures with associated concepts of the semiotic system such as mechanical bonds [198], transition states and HOMO-LUMO gaps. Variation of reaction conditions have also expanded the conventional chemical possibilities of reactive partners such as amines and carboxylic acids [199] and extreme regimes of pressure and temperature [200] have led to compounds with unconventional stoichiometries [201]. The very identification of substances has evolved, as ideas on the essential properties of substances have changed over the history.[27] Initially, the identity of a substance was based on the material it was extracted from [71], then on composition and chemical analysis and synthesis; with physical, chemical and secondary properties playing a major role [155]. Then, the advent of structural theory turned molecular structure into the essential property of substances by assuming a one-to-one relationship between substances and structures [26,202]. This move to structures occurred hand in hand with the means to determine them, initially through chemical means, and after the 1950s, as discussed through spectroscopic techniques [155, 194]. Today, we talk about substances that may not be stable under ambient conditions, whose transient existences are determined by spectroscopic means under very particular conditions (Fig. 3.5).[28]

Another instance of the evolution of the material system is through the evolution of the chemical laboratory [204], which includes the evolution of the technologies to separate and explore the behaviours of substances and to produce new ones (Fig. 3.5). Chemists have moved from furnaces and distillation apparatus to Bunsen burners, spectroscopic, chromatographic and robotic devices. These technologies have also influenced the scales and speeds of work, ranging from human to micro scales and from days to milli, femto and attoseconds as currently allowed by lab-on-a-chip, microfluidic, high-throughput techniques [205–208], as well as femto and attochemistry methods [209, 210]. Today

few or no further new substances, therefore reducing the number of new substances reached by synthetic methods. The question that arises is what happened with the synthetic activity after the 1950s, as chemists, now with more time not devoted to structural determinations [156], could spend it on further synthetic activities. More research is needed for more insight on the balance between synthesis and extraction brought up by the introduction of spectroscopic tools. Nevertheless, as discussed in footnote 5, after the 1950s there was a surge in the per capita number of new chemicals reported, which seems to indicate that spectroscopic methods actually favoured the report of new natural products, typically resulting from extractions [195]. An interesting reference analysing the evolution of this chemistry subdiscipline is [196].

[26] A finer temporal resolution of processes leads to the reactive intermediates, such as radicals, diradicals, carbenes, carbocations, carbanions, and zwitterions. Also, entropic intermediates are now added [197].

[27] Chemists have always sought for a reduced set of properties allowing them to apply the Leibnizian principle of the identity of the indiscernibles, where any two substances are said to be the "same" insofar as they hold the same essential properties [155].

[28] An example is Na_2He, observed at pressures greater than 113 GPa and detected using synchroton X-ray diffraction and Raman spectroscopy [203].

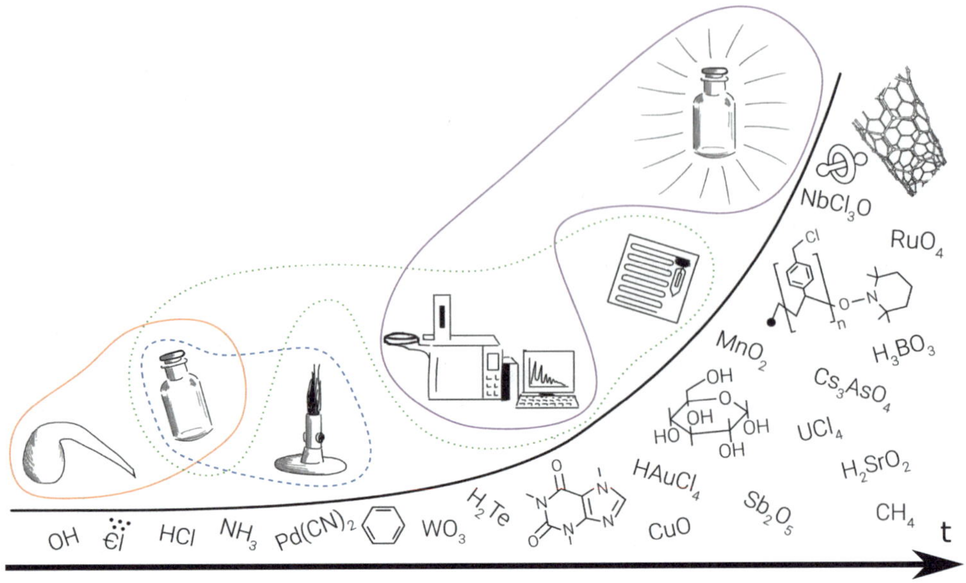

Fig. 3.5 Evolution of the material system of chemical knowledge. It is represented the exponential growth of the chemical space by including several chemical notations for different substances. Above the exponential curve, the evolution of chemical instrumentation is illustrated, from glassware such as a retort and a flask; and separation apparatus such as a gas chromatographer, to Bunsen burners and microfluidic technologies. The transient character of contemporary chemical species is represented on the right with the glass bottle with emerging lines. Subsets (or hyperedges) indicate internal relationships among material objects: eighteenth- and nineteenth-century glassware (left-hand-side subset), mid nineteenth-century laboratory instrumentation (dashed-line subset), storage and substance separation (dotted-line subset) and modern instrumentation for separation and "storage" of chemical species (right-hand-side subset)

many chemical laboratories are very sterilised and robotised places, where the volume of matter under study is surpassed, by several orders of magnitude, by the apparatus used to investigate it. The evolution of the laboratory also shows the interaction between the social and the material systems, where alchemical, metallurgical, agricultural and pharmaceutical interests shaped laboratories and where the rise of chemistry as an academic and professional discipline influenced the size and purposes of these spaces [204].

4 Interactions Between Systems, Their Differences and Environments

Abstract

We present some instances of binary interactions among the social, material and semiotic systems of chemical knowledge. We highlight the relevance of the different temporalities of each system for the purposes of modelling the evolution of chemical knowledge. Finally, we discuss the relationship of chemical knowledge with other kinds of scientific knowledge.

As expected, the dynamics of the three systems of chemical knowledge are not independent of each other.[1] Semiotic conventions express and organise knowledge about the material system, and they are invented, codified and transmitted by social actors. The surprising results of new discoveries in turn are understood within new semiotic principles and rules, and they bring new social actors to the fore. Semiotic principles guide the exploration of the chemical space. And chemistry, like other fields, is of course socially organised. We shall now discuss some more concrete examples of such interactions.

4.1 Material-Semiotic Interactions

Besides the discussed vibrant dynamics between the material and the semiotic systems, some particular examples of this interaction are Lavoisier's chemical nomenclature, Berzelius' symbolic notation for elements and compounds' formulae [166]; and Friedrich August Kekulé's (1829–1896) benezenoid hexagons [211].

[1] See footnote 16 in Chap. 1.

J. Jost, G. Restrepo, *The Evolution of Chemical Knowledge*, Wissenschaft und Philosophie – Science and Philosophy – Sciences et Philosophie,
https://doi.org/10.1007/978-3-031-10094-9_4

Lavoisier's nomenclature was motivated by several changes in the material system. For example, the adoption of more precise laboratory instruments and quantitative methods [35, 212]; the chemistry of gases, which added a third dimension to the classical two based on solids and liquids [29]; and the size of the chemical space that made evident the problems of the methods to name substances. Lavoisier's nomenclature changed the way of communicating chemistry, not only because it introduced new signs (names) for old and new semiotic objects (substances), but also, and even more importantly, because interpretants (chemists) changed the way of addressing and understanding chemical phenomena.[2] This nomenclature was in turn influenced by the material system. The name of a compound should incorporate the class of chemicals leading to it, for example bases and acids to salts. The name, in turn, would lead to infer the possible reactions of a substance based on such classification. Hence, Lavoisier's nomenclature aimed at encoding chemical knowledge, analytic and synthetic, in its substance names. In the case of elements, names should reflect some of their properties. A famous case is oxygen, as Lavoisier erroneously believed that all acids contained oxygen (see Sect. 2.2.2). A further example involves suffixes for naming 86 of the currently known elements, whose names in English use the suffix *-ium* or *-um*, which is derived from the material system, as these suffixes refer, in Latin, to the metallic character of those substances [214]. Metallicity, or better its absence, is also part of the names of non-metallic elements through the *-on* or *-ine* suffix (except for helium and selenium).[3]

Berzelius' notation constitutes another instance of interaction between the semiotic and the material system of chemical knowledge. The size and diversity of the chemical space provided, in the early years of the nineteenth century, a considerable number of inorganic substances and a growing number of organic ones [4, 92, 216], which led to several important quantitative results giving place to stoichiometric laws. Organic compounds turned out to be problematic, as several of them had the same composition, challenging composition as an identification feature [71]. Berzelius' notation, based on arrangements (text strings) of Roman alphabetic characters and Arabic numbers, solved the problem by representing composition and the binary constitution of chemical compounds. This latter concept Berzelius borrowed from inorganic chemistry and extended to organic

[2] This was a clear aim of Lavoisier's nomenclature. As he stated it in the preface of "Elements of chemistry" [213], p. xix, xv (original text in French in [165], p. vi, vii):

> [...] as ideas are preserved and communicated by means of words, it necessarily follows that we cannot improve the language of any science without at the same time improving the science itself; neither can we, on the other hand, improve a science, without improving the language or nomenclature which belongs to it.

[3] Interestingly, nihonium, moscovium, tennessine and oganesson, which are the names of the most recently synthesised elements, indicate metallic properties for the first two, while non-metallic for the others. Nevertheless, these properties are experimentally difficult to measure and theoretical models are used instead [39], but these results do not always agree [215].

substances. By indicating substance constitution, Berzelius encoded the past, present and future of the substance, continuing Lavoisier's tradition, as his notation, giving place to constitution formulae, indicated those substances producing the substance in question and those obtained by its chemical analysis [71]. Hence, the material-semiotic relationships motivated by Berzelius' notation become clear. This notation and its associated formulae introduced a powerful semiotic tool, a paper tool according to Klein [71], that changed the way of understanding chemical substances, of exploring the chemical space and which gave place to further semiotic advances such as type, structural and stereochemical formulae [71]. These formulae, in particular structural ones, constitute one of the essential communication objects of chemistry [26, 31].

Kekulé's benzene hexagon constitutes a further case of the material-semiotic relationships. It arose in 1865/66 as the theoretical solution to the mounting evidence that aromatic substances were different from the others [217, 218],[4] especially given their high chemical stability when compared with others of similar composition. These substances, with benzene as its most representative compound, therefore challenged the traditional chemical classification of substances. Benzene constitution indicated an unsaturated compound. However, it did not undergo reactions of this class of compounds [219]. Moreover, the recurrent number of benzene derivatives[5] suggested to Kekulé a regular hexagon of carbons as its molecular structure. Such hexagon with alternating bonds, actually a cyclohexatriene, was replaced by Kekulé in 1872 for a hexagon where single and double bonds rapidly exchange places leading to a highly symmetric structure differentiating it from typical unsaturated compounds [219, 220]. Kekulé's hexagon and its alternating bonds became a semiotic token of the chemical language, a rich source for further theoretical work and above all, a guiding principle to understand and figure out new reaches of the chemical space.

More recent instances of the semiotic-material interaction include Eugene A. Markush' (1888–1968) structures [221] indicating backbone commonality of molecular structures; the adoption of "Pol" or a filled dot or sphere as a quasi-elemental symbol, as H and C are, representing resins in Robert Bruce Merrifield's (1921–2006) solid phase synthesis [222, 223]; and James Fraser Stoddart's (1942–) representations of mechanically interlocked molecules.

Markush' structures, introduced in 1923/24 [224], have a clear semiotic aim: developing the most general sign (molecular structure) encompassing the largest number of chemical compounds (semiotic objects). Its origin, besides being justified by the material system of chemistry, is also related to the social and the very semiotic system. It is social

[4] As Alan J. Rocke (1948–) pointed out, chemists could not address the problem of aromatic substances before because the chemical space was not populated with many members of these substances. This claim is confirmed by our results analysing the chemical space from 1800 to 1869 [216].

[5] That is one, three, three, three, one and one mono, di, tri, tetra, penta and hexa-substituted derivatives, respectively.

because these structures arose from needs of the chemical industry and of the patent system, where patent applicants typically aim at covering large numbers of substances and processes in their applications. It is also a semiotic case as it is caused by the internal dynamics of chemical communication. Markush' structures became new semiotic objects widely used by the community as still evidenced in the chemical literature, for instance in organic chemistry publications, patent applications [224] and in the amount of chemoinformatics and mathematical chemistry research they have generated [225].

More recently. Solid phase synthesis introduced "Pol" or a filled dot as a new semiotic object. This was motivated by particular advances in the material system starting in the 1950s with Merrifield's synthesis of peptides, which is based on a solid support (typically a resin) that bonds molecules in a step-by-step manner through chemical (de)protecting reactions.[6] Solid support synthesis currently goes beyond peptide synthesis, covering DNA, RNA and other substances based on strings of molecules, plus small molecules such as heterocycles [226]. Therefore, it is the source of many novel substances of the chemical space. As the use of the resins in these syntheses became common place or black boxes in Latourian terms [227], instead of often repeating the molecular structure of the resin, chemists began to replace it by "Pol," indicating a polymer (typically polystyrene cross-linked with divinylbenzene [226]) or by a filled dot or even a sphere. The semiotic load in these signs emphasises two aspects of the resin: "Pol" focuses on the chemical nature of the resin, that is, its chemical classification, while either the filled dot or the sphere refer to the macroscopic view of the resin, normally resin beads of 100–200 and 200–400 mesh (75–150 and 35–75 microns, respectively).[7]

The fourth example of the material-semiotic relationship we discuss is given by Stoddart's representations of mechanically interlocked molecules. At the nanometer scale, it becomes spatially possible to arrange the component parts of these molecules in such a manner that breaking the structure can only be afforded by breaking conventional chemical bonds. Hence, what actually "holds together" the molecular components of these structures is a new kind of chemical bond, the mechanical bond [198]. Figure 4.1 shows a mechanically interlocked molecule with two components (red and blue). When covalent bonds, lines in this representation, are endowed with actual interatomic distances, the impossibility of sliding the red component out of the blue one becomes clear. Note that benzenic rings in Fig. 4.1 do not correspond to the typical ones depicting one of the two resonance structures as the cyclohexatriene structure depicted on the left of the Markush' structure in Fig. 4.1. Mechanically interlocked molecules are depicted with Kekuléan hexagons incorporating solid internal circles (full moons) that denote aromaticity in π-conjugated ring systems. The semiotic load here is shifted from typical relations between

[6] Interestingly, solid phase synthesis arose as a solution to a problem in the material system, namely that the isolation of intermediates was the most time-consuming step in traditional pre-1950s peptide synthesis [226].

[7] More on the conventions of solid state synthesis is found in [226].

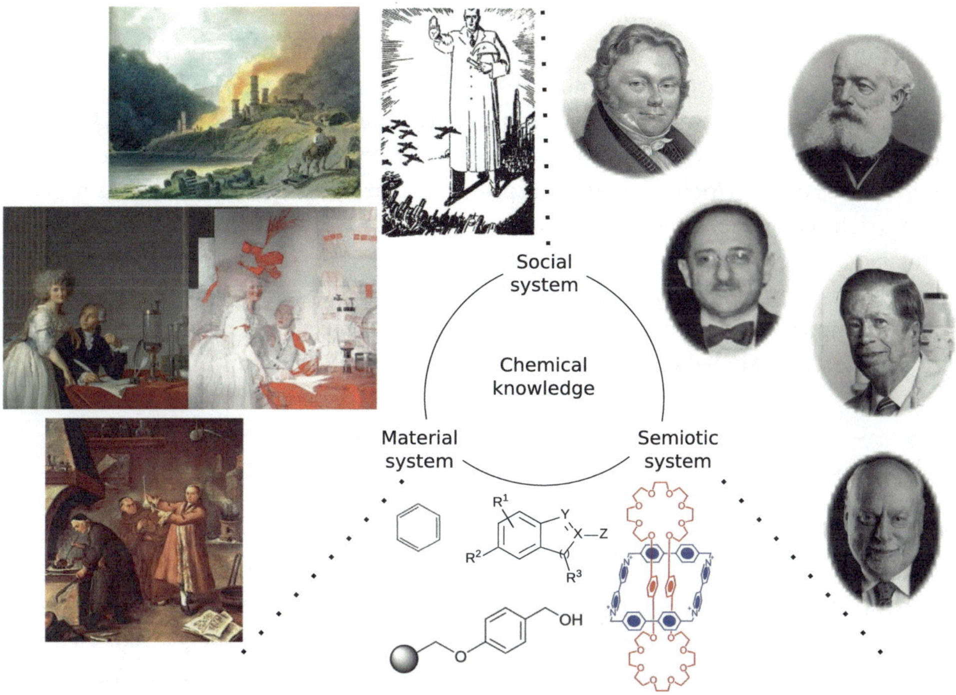

Fig. 4.1 Binary relationships between the systems of chemical knowledge. At the bottom some instances of the material-semiotic relationships are depicted: the iconic benzene ring; a Markush' structure [229] where R_x indicates substituents, $()_n$ an addition of n carbons to the cycle and X, Y, Z any chemical element, typically N or O; the Wang resin' structure with the filled dot; and a donor–acceptor ring-in-ring complex (mechanically interlocked molecule with π-donor and π-acceptors macrocycles in red and blue, respectively) with full moons (reprinted with permission from Fig. 24 in [226]. Copyright 2021 American Chemical Society). On the left, instances of the relationship between the material and social systems. Bottom left image, "The Alchemists," c. 1757, oil on canvas, 61 × 50 cm (Galery Ca' Rezzonico Venice) by Pietro Longhi (1702–1785) (reproduced from Fig. 6 in [230]). Picture above, left, detail of portrait of Antoine Laurent Lavoisier (1743–1794) and Marie-Anne Lavoisier (Marie-Anne Pierrette Paulze) (1758–1836) of 1788. Oil on canvas, 259.7 × 194.8 cm by Jaques-Louis David (1748–1825), reproduced from graphical abstract of [231]. Right, detail of combined Pb (white) and Hg (red) distribution maps for selected areas of the portrait [231]. Picture above, "Iron Works, Colebrook Dale" (1805) by William Picket for Philippe Jacques de Loutherbourg (reproduced from Fig. 13 in [230]). Black and white drawing above: "Chemist protecting society in 'Security'," from [232] and reproduced from Fig. 5 in [233]). On the right five scientists are presented (from left to right, top-down) as representatives of the social-semiotic relationship: Jöns Jakob Berzelius (1779–1848), Friedrich August Kekulé (1829–1896), Eugen A. Markush (1888–1968), Robert Bruce Merrifield (1921–2006) and James Fraser Stoddart (1942–)

atoms such as in resonance structures.[8] In this latter case the aim is to depict delocalised π electrons between carbon atoms, while in Stoddart's representations the aim is to indicate interactions between component parts of the molecules, for instance the red and blue macrocycles shown in the mechanically interlocked molecule at the bottom of Fig. 4.1. These macrocycles are coloured according to their molecular connectivity and to their electron-poor/rich features. In Fig. 4.1, the red macrocycle corresponds to the π-donor (rich) component, while the blue macrocycle to the π-acceptor or electronic poor component. Mechanically interlocked molecules arose from the expansion of the chemical space, as part of the material system of chemical knowledge, and motivated new forms of representation and the introduction of a new chemical concept: the mechanical bond. This is a clear relationship between the material and the semiotic systems of chemistry.[9]

4.2 Social-Semiotic Interactions

In the previous section we presented some examples of the material-semiotic interaction of the systems of chemical knowledge, which were championed by leading chemists in different periods of the history of chemistry. These chemists conducted their researches and afforded semiotic changes based on the social contexts in which they acted. Therefore, they can be regarded as exemplars of the social-semiotic interaction of the systems of chemical knowledge. In particular, we discussed how Lavoisier, Berzelius, Kekulé, Markush, Merrifield and Stoddart influenced chemical knowledge. Some portraits of these chemists are shown in Fig. 4.1. More explicit examples of scientists leaving their marks upon semiotic tokens of chemistry include Nicolaus Copernicus (1473–1543), Vittorio Algarotti (1533–1604), Johann Rudolf Glauber (1604–1670) and Wilhelm Homberg (1652–1715), who marked chemical language with copernicium, powder of Algaroth (antimony oxychloride), Glauber' spirit of nitre (fuming nitric acid) and Homberg' salt (boric acid).[10]

[8] Or even Clar structures [198].

[9] A further case of interaction between the semiotic and the material systems is given by the ribbon diagrams to represent the three-dimensional structure of proteins, introduced by Jane S. Richardson (1941–) in 1980 [228].

An interesting example of the influence of the material system upon the semiotic one is given by Leicester [35]. In Sumerian, *udun* means melting furnace, which became *utunu* in Babylonian and *tannur* in Arabic. By adding the Arabic article *al*, it passed to Latin as *Althanor*, a famous furnace used by alchemists.

[10] More recent examples include Johan Gadolin (1760–1852) (gadolinium), Alfred Bernhard Nobel (1833–1896) (nobelium), Dmitri Ivanovich Mendeleev (1834–1907) (mendelevium), Wilhelm Conrad Röntgen (1845–1923) (roentgenium), Marie Salomea Skłodowska Curie (1867–1934) and Pierre Curie (1859–1906) (curium), Ernest Rutherford (1871–1937) (rutherfordium), Lise Meitner (1878–1968) (meitnerium), Albert Einstein (1879–1955) (einsteinium), Niels Henrik David Bohr (1885–1962) (bohrium), Enrico Fermi (1901–1954) (fermium), Ernest Orlando Lawrence (1901–

The history of chemistry offers further cases of the social-semiotic interaction. A famous case is the cryptic writing style of some alchemists to avoid spreading the knowledge of some particular preparations to unauthorised people and the influence of oriental mysticism in alchemy, which added many new philosophical speculative terms to the already confused terminology [35]. It is noteworthy how political changes have left their mark in the semiotics of chemistry as evidenced in the several terms used still today, which include Egyptian, Syrian, Persian, Arabic and European traditions [35, 118].[11] A more recent instance of the social-semiotic interaction involves the way chemists report chemical reactions, which has evolved from lengthy and detailed descriptions of analytical and synthetic experiments to the current ones where chromatographic and spectroscopic signals are combined with brief descriptions of substrates and targets often without mentioning side products.[12]

As the social system also includes institutions and higher order organisations such as states and countries, the binary interaction between the social and the semiotic system includes cases of these forms of organisation influencing the semiotics of chemistry, as well as this latter influencing higher order organisation forms. Chemical elements constitute a clear case of the former [235, 236], for instance with elements named after particular villages or cities, countries, continents or even rivers [237]. These include ytterbium, darmstadtium, francium, americium and rhenium.[13] Element names also depict political tensions, for instance when Marie Salomea Skłodowska Curie (1867–1934) and Pierre Curie (1859–1906) discovered a new element and named it polonium in honour of Marie's homeland. This was a clear political movement, as there was no Poland on

1958) (lawrencium), Glenn Theodore Seaborg (1912–1999) (seaborgium), Georgy Nikolayevich Flyorov (also spelled Flerov) (1913–1990) (flerovium) and Yuri Tsolakovich Oganessian (1933–) (oganesson).

[11] An excellent account of the social influence upon the systematic nomenclature of organic chemistry between 1889–1940 is found in [234] and in Evan Hepler-Smith's (1984–) coming book.

[12] Therefore, a large proportion of the chemical information about reactions involves non balanced reactions and reactions that are not even full. A reaction is full when all substrates and products are reported, and it is balanced when the number of atoms of each participating element is preserved during the reaction. Of course, actual chemical reactions are always full and balanced, but the reports often suppress side products like water or carbon dioxide that are considered as being of no interest. This form of communication is the consequence of tacit agreements in the interior of the social system of chemistry.

[13] A more complete list includes ruthenium (named after Ruthenia, the Latin word for Russia), yttrium, terbium, erbium, ytterbium (after the village of Ytterby in Sweden), hafnium (after Hafnia, the Latin name for Copenhagen), rhenium (after the river Rhine in Europe), hassium (after the German state of Hesse, or Hassia in Latin), nihonium (after the common Japanese name for Japan), livermorium (after Livermore, an American city), holmium (after Holmia, the Latin name for the city of Stockholm), thulium (after Thule, an Ancient Greek place name associated with Scandinavia or Iceland), lutetium (after Lutetia, the ancient Roman name for Paris), dubnium (after the town of Dubna in Russia), gallium (after Gallia, the Latin word for France), germanium, scandium, europium, moscovium, polonium, tennessine, berkelium and californium.

any of the maps at that time. Marie Curie's intention was bringing the world's attention to Poland's lack of independence [236].[14] A further example includes the nationalistic overtones pushing to rename iconic concepts in chemistry such as the periodic system of chemical elements for Mendeleev's periodic system, as submitted in 2013 to the International Union of Pure and Applied Chemistry (IUPAC) and more recently as a general motion to UNESCO [39].[15]

The influence of the semiotic system of chemical knowledge upon the social system is at the core of chemistry. The semiotic system contains the logical structure of chemistry, where concepts are related and theories are built up. Over the history, chemists have spread their knowledge and its structure mainly through manuscripts, books and handbooks, which have influenced the way of teaching the discipline. This is evident, for instance, in the chemical knowledge passed through apothecary or medical schools, as well as through metallurgy and mineralogy departments. The historical expansion of chemical knowledge, with its disciplinary branching, has been mirrored at the social level in all times and it is particularly evident in the twentieth-century academic division of chemistry, where biochemistry became a new discipline and this led to reorganising universities and

[14] We mentioned before in this section the case of element mendelevium as a case where a scientist left his mark upon a semiotic token of chemistry. This element was actually named after the Russian scientist not only because of Mendeleev's glory but also because of the political tensions of the Cold War. According to Juris Meija [236]:

> [I]n 1955, Glenn Seaborg, Albert Ghiorso and colleagues at Berkeley created element 101 and named it mendelevium. With tensions between the USA and USSR sky-high, many criticized this move. However, years later both Ghiorso and Seaborg recalled the positive impact the naming of mendelevium had on US and Soviet relations at that time. Ghiorso remembers the physicist Moïse Haïssinsky telling him during the 1955 Atoms for Peace Conference in Geneva that the naming of an element in honour of a Russian scientist had probably done more good for international relations than anything US Secretary of State John Foster Dulles had ever done. Vice President Nixon even made use of the mendelevium story during his visit to the Soviet Union in 1959, thanks to a letter from his long-time friend Seaborg.

[15] This latter was proposed during a scientific meeting, the twenty-first Mendeleev Congress in St. Petersburg, 9–13 September 2019 [39]. Still at the level of chemical elements, an interesting paper highlighting the political context of the synthesis, and above all the confirmation of those syntheses, is [238]. There it is mentioned how after the report of the first transfermium elements in the 1950s, largely by the American Lawrence Berkeley Laboratory and the Russian Joint Institute for Nuclear Research, tensions started to arise. "After each new element was claimed to have been discovered by one lab, the claim was contested by the other. The International Union of Pure and Applied Chemistry (IUPAC) and the International Union of Pure and Applied Physics (IUPAP) formed a joint working group to end the controversies, the Joint Neutral Group (JNG). When that group failed to resolve the discovery disputes, another was formed, the Transfermium Working Group (TWG). Neutrality was a value important to both groups, giving them the credibility necessary to act as mediators" [238].

research institutes. Current disciplinary branching involves material sciences, to which the social structure of chemistry is adapting.

4.3 Material-Social Interactions

The relationship between the material and the social systems is evident, for example in the sixteenth-century improvement of distillation methods which led to explore the chemical space with very reactive substances such as mineral acids and also organic acids [35], which along with alcohols expanded the possibilities of chemical reactions in solutions.[16] This caused a surge in the demand of these substances which led to the social organisation around a primitive chemical industry [35, 239].[17] Another instance of the social-material interaction are the pharmaceutical companies created in the subsequent centuries from the chemistry of dyes and the chemical knowledge of apothecaries [56, 239].[18] Likewise, the nineteenth-century deluge of organic chemicals influenced the creation of new academic programs and of specialised research laboratories [204] (Fig. 4.1).[19] The social system also motivated developments in the material system, for example through Lavoisier and his nomenclature or through Berzelius and his formulae, which became tools to enlarge the chemical space.[20] Other examples include the anthropogenic fashion in which the chemical space has been explored, where preferences for starting materials [4] and reactions [67] bias the possibilities of exploration [4, 69], leaving some regions

[16] In [4], we show the still important role of acetic, hydrochloric, nitric and sulfuric acids as some of the most used substrates in the nineteenth century. Ethanol was also a widely used substance in this period [4].

[17] As noted in Chap. 1, the material output of chemistry shapes the present world. Although this was not yet a fact in the sixteenth century, Glauber, an influential chemist of this period, saw the benefits of chemistry for the society. In 1656 he published his *Teutschlands Wohlfahrt* (The prosperity of Germany), where he applied chemical principles in his recommendations for making Germany an economically self-sufficient territory [35].

[18] It has been claimed that the overthrow of the feudal system was strongly attached to the development of gunpowder as an instrument of warfare [35]. This constitutes an important interaction of the material and social systems of chemical knowledge.

[19] The growth of the chemical space in the sixteenth century brought better understanding of acids, bases and salts, as well as the introduction of competing theories trying to make sense of such space. This, in turn, led to the publication of several influential chemistry textbooks, as *Cours de chimie* (1675), by Nicolas Lémery (1645–1715) [35], which caused a widespread interest in chemistry bringing more human power to the laboratory. Chemistry textbooks are also a powerful semiotic device to spread chemical knowledge, as we have noted before.

[20] The historical association of the experimental research with technology (especially important in chemistry, as discussed before), has the consequence that social aspects involving qualified personnel as well as specialised equipment, restrict access to certain knowledge to only those research groups with the means and infrastructures to afford them. This constitutes another facet of the influence the social system has on the material system.

of the space essentially empty, while heavily populating others by similar compounds [4, 196, 240, 241].[21] A further case of the social and material interaction is the recent rise of prophetic substances, that is compounds discussed in the literature that have not been synthesised or extracted [242].[22]

In the material and social relationship, the encyclopedists of the thirteenth century played a major role in systematically compiling, organising and curating the amount of chemical knowledge produced from Greek and Hellenistic times to Chinese and Arabic alchemy. They played an instrumental role in reintroducing the seeds of modern chemistry to the West. Chief names among them include Batholomaeus Anglicus (before 1203–1272) with his *De proprietatibus rerum* (On the properties of things) [35, 246]; Vincent of Beauvais (c. 1190–c. 1264), who contributed the *Speculum Maius* (Greater mirror) [35, 247][23] and Albertus Magnus (1193–1280), who wrote the *De mineralibus* (On minerals) [35]. The importance and practical details of distillation was popularised by Hieronymus Brunschwygk's (c. 1450–1513) *Liber de arte distillandi de simplicibus* (1500) and Great Book of Distillation (1512) [35]. Likewise, Ulrich Rülein von Calw (1465–1523) gathered knowledge on metallurgical and mineralogical chemistry in his *Eyn wohlgeordnet und nützlich büchlein, wie man bergwerk suchen und finden soll* (1505) and *Ein nützlich Bergbüchlin* (1527). Other important sources indicating the advance of the material system in the sixteenth century are *De la pirotechnia* (1540), by Vannucio Biringuccio (c. 1480-c. 1539) and *De re metallica* (1556), by Georg Bauer (1494–1555), known as Agricola for the Latin form of his surname [35]. At the turn to the seventeenth century, Andreas Libau (1550–1616), or Libavius in the Latin form, included in a single book (*Alchemia*) all rational and logical chemical knowledge that had been scattered in the alchemical, pharmaceutical and metallurgical literature [35] (p. 98,99). Glauber also collected and organised information on chemical instrumentation in his *Furni Novi Philosophici* (New philosophical furnaces) of 1648–1650 [35] as well as on iatrochemical recipies in *Pharmacopoeia Spagirica*. These publications contributed to provide a whole

[21] It has been claimed that a possible way to explore the chemical space in an unbiased manner is through algorithms capable of analysing the available chemical space, which when coupled to robotic sensor arrays will unveil new regions of the space [69].

[22] The effects of the social system upon the material system of chemistry are also evident in the lack of experimentation in ancient Greek science, which contrasts with the experimental traditions in preceding Mesopotamian and Egyptian cultures. Ancient Greeks believed that "work with hands was for the slave," while "work with the mind, for the free citizen" [35]. Nevertheless, in an era of social changes, there were some instances of Greek experimentation, as Empedocles' (ca. 495-435 BC) *klepsydra* experiments proving the materiality of air [243, 244]. In the Hellenistic era, experiments became more widespread and more systematic, although this was not a very productive period for chemistry, as the scientists of that period were more interested in mechanical devices than in chemical transformations [245].

[23] Central for chemistry was his *Speculum Naturale* (Mirror of nature), in particular books v to xiv [247]. The *Speculum Naturale* was part of *Speculum Maius*.

picture of the chemical knowledge of their times and influenced the way chemistry practitioners expanded the material system.

Crucial for the unfolding of chemical knowledge in the eighteenth century were the multiple efforts to collect information to elaborate on the nature of chemical reactions and the forces holding compounds together. Leading figures of this chemical programme were Geoffroy, who in 1778 set the stage for collecting and systematising information on chemical reactions through affinity tables (Fig. 3.2). These semiotic objects were enlarged over the century as the chemical space expanded. Chief figures in this movement include Jean Grosse (?–1744), Christlieb Ehregott Gellert (1713–1795), Jean Philippe de Limbourg (1726–1811) and Bergman,[24] with his large 1775 and 1783 tables [157].[25]

More recent examples of the material-social interaction include Leopold Gmelin (1788–1853) and Friedrich Konrad Beilstein (1838–1906), who were also central as they spread and systematised chemical knowledge up to the nineteenth century through their colossal respective Handbooks of inorganic and organic chemistry. As the examples discussed before, these handbooks guided the expansion of the material system by providing a systematised corpus of what was already achieved and by indicating gaps in the material system to be filled in. A further material-social example is the role chemical societies have had in the evolution of chemical knowledge. Some of them are the *Deutsche Chemische Gesellschaft*, merged with the *Verein Deutscher Chemiker* into the current *Gesellschaft Deutscher Chemiker* at the dawn of the twentieth century [55]; and the American Chemical Society created at the end of the nineteenth century. As Anglicus, Biringuccio, Bergman and Gmelin, these societies are instrumental for spreading chemical knowledge through their communication channels, including scientific journals and meetings.

Du Pont and the several large scale laboratories for nuclear research, which resulted from post-WW2 social changes in the scientific landscape are instances of twentieth-century material-social interactions. The research possibilities at Du Pont after WW2 were unique, as they involved very qualified chemists with access to the most advanced technology and with enough financial support to venture into new areas of industrial and fundamental research leading to new products, several of them enormously successful in the market [248].[26] This unique research and development environment expanded the chemical space in different ways.[27] Another region of the chemical space that was expanded as a consequence of the synergies at Du Pont was that of fluorinated

[24] See Sect. 1.3.

[25] A detailed analysis of chemical affinity and affinity tables is found in [157].

[26] They include ultra-high molecular weight polyethylene used in hip and knee replacement arthroplasty [249]; urea and uracil compounds used as herbicides, along with sulfonylurea [250]; and potassium titanyl phosphate, a nonlinear optical material [251].

[27] A further instance of this expansion of the chemical space, initiated at Du Pont, and motivated by an actor of the social system is the case of Earl Muetterties' (1927–1984). He contributed to a large extent to boron hydride chemistry through the synthesis of the polyhedral borane anions

compounds. Although these substances have a long tradition in chemistry, research on organofluorinated compounds skyrocketed after WW2. Du Pont participation in the Manhattan Project provided its chemists and engineers with the needed expertise to handle these new substances [253]. Likewise, Du Pont researchers, especially Owen Wright Webster (1929–2018) and Theodore L. Cairns (1914–1994) contributed to the synthesis of new polymers [254, 255], continuing the greatest achievements of the company in this area, which began with the synthesis of nylon-6,6 and polychloroprene (neoprene) in the early 1930s.[28]

The Lawrence Livermore National Laboratory (USA), the Joint Institute for Nuclear Research (Russia), the Helmholtz Center for Heavy Ion Research (Germany) and the RIKEN Nishina Center for Accelerator-based Science (Japan) constitute twentieth- and twenty-first-century cases of the interplay between the social and the material system of chemistry. World Wars showed the central role of scientific knowledge for military aims and motivated drastic changes in the social system of science, including that of chemistry. This involved a surge in interdisciplinary collaboration, including more internationalism; more research budget for large scale projects, with leading roles of national governments and further specialisation [256, 257]. As discussed in the previous paragraph, Du Pont benefited to a large extent from this military research and the infrastructure it brought, leading to several new types of compounds. But the material system was also affected at a most fundamental level by the post-WW2 military research, namely at the atomic and nuclear one through the synthesis of new chemical elements, where the American, Russian, German and Japanese institutions mentioned before played a leading role.[29] The synthetic elements not only enlarged the alphabet of chemistry, but triggered novel concepts, methodologies and instrumentation to treat atomic species, which in most of the cases last much less than fractions of seconds.

4.4 Lavoisier: Bridging Systems of Chemical Knowledge

An interesting and sad event of the history of chemistry, which goes beyond the binary interactions among systems above described is the execution of Lavoisier during the Reign of Terror of the French Revolution, which perturbed all the different systems of chemical knowledge. Lavoisier was a very influential (perhaps the most influential) chemist of the

$B_nH_n^{2-}$ [252], which expanded an important region of the inorganic chemistry space. We have shown elsewhere how boron chemical space grew fast after 1950 [4].

[28] An interesting account of how different social, scientific and political factors played a major role in the success of Du Pont is found in [248], which, as we have seen here, turn out to be important social factors influencing the material system of chemistry.

[29] These institutions have been involved in the synthesis of 13 heavy elements: nobelium, bohrium, hassium, meitnerium, darmstadtium, roentgenium, copernicium, nihonium, flerovium, moscovium, livermorium, tennessine and oganesson.

eighteenth century and as such played an important role in the social system of chemistry of his time. He was also active in the external environment of chemistry (discussed in Sect. 4.7), as evident in his economic suggestions, which might have alleviated the social distress of the pre-revolution times if they had been accepted [35]. He was also part of the external social environment of chemical knowledge through his belonging to the *Ferme générale*, which ultimately led him to the guillotine.[30] A dramatic evidence of Lavoisier' social interactions is observed through the hidden layers of the most famous portrait of the French chemist and his wife, Marie-Anne Pierrette Paulze (Marie-Anne Lavoisier) (1758–1836), part of it shown in Fig. 4.1. The portrait was painted by Jaques-Louis David (1748–1825) in 1788 and depicts the couple with some chemical instruments, some on the table in front of Lavoisier, others, lying on the floor, apparently pushed aside by Lavoisier's right leg. Lavoisier is presented seated writing on a table and gazing at the central figure of David's portrait: Marie-Anne, who is leaning on her husband's shoulder and looking at the painter. A recent analysis of the portrait at the New York's Metropolitan Museum of Art revealed that the original portrait depicted something different than the scientific interests of the couple [231],[31] it actually showed Lavoisiers' social status. The original portrait did not contain any scientific instrument, it rather depicted Marie-Anne with a "huge, ribboned black and red hat festooned with flowers in a style fashionable in late 1787, characteristic of the conspicuous consumption of the ruling classes" [259]. The researchers who analysed the painting believe the additions were painted over between 1787–1788 presumably to reassure the image of the couple in a France at the eve of the revolution [231].[32] Further research is needed to determine whether the last-minute changes obey David's own initiative of whether they resulted from a desperate request from the couple. At any rate, the story of David's painting is the dramatic story of a scientist embedded in his social system.

Lavoisier launched along with Louis-Bernard Guyton de Morveau (1737–1816), Gaspard Monge (1746–1818), Claude Luois Berthollet (1748–1822), Philippe Friedrich Dietrich (1748–1793), Jean-Henri Hassenfratz (1755–1827) and Pierre-Auguste Adet (1763–1834) the influential *Annales de chimie*, which was one of the first scientific journals entirely devoted to chemistry. This was a further addition of Lavoisier to the semiotic system of chemistry, which complemented his contributions to chemical nomenclature and his new setting for chemistry, which constituted a powerful paper

[30] The *Ferme générale* was an organisation of financiers that leased from the government the right to collect taxes on many essentials of commerce. Revolutionaries considered members of the *Ferme générale* and of the aristocracy responsible for many of the pre-revolutionary social distresses [35].

[31] Marie-Anne Lavoisier was central for her husband' scientific career. She was his chief collaborator and laboratory assistant. She also translated several English texts into French to ease her husband's research. Likewise, she drew many sketches and carved engravings of chemical instrumentation, which were used by her husband in his publications [29] (pp. 92–93), [258].

[32] A sketch of the original painting (line drawing), produced by tracing information obtained by infrared reflectography and X-ray fluorescence is found in [260].

tool [71] to expand the chemical space, therefore the material system of chemistry.[33] A further addition to this system was the use and popularisation he made of the most advanced chemical instrumentation of his time. Lavoisier is therefore an example of a chemist contributing to chemical knowledge by influencing its various systems and its environment, who was finally silenced by this environment.

4.5 Interactions Between the Material, the Semiotic and the Social System of Chemistry

In Chap. 1, we have already argued that chemical knowledge can be understood and consequently should be conceptualised only from the simultaneous perspective of the three systems, the material, the semiotic and the social one. Chemical knowledge is about material properties, not only of substances and reactions, but also about chemical methods and practices. It is formulated, encoded and expressed with the chemical symbols and concepts created in the semiotic system and therefore reflects the scientific theories of its time. It cannot be seen as independent of theory, but each piece of chemical knowledge is embedded into a large theoretical web. From an abstract perspective, we represent the knowledge of substances and reactions as a hypergraph that is growing in time. This abstract concept will be elaborated upon in the second part of this book. Moreover, chemical knowledge is embedded into and transmitted within a social system. And conversely, chemical knowledge not only consists of intrinsic properties of reactions and substances, but also contains information about their uses and affordances outside the science of chemistry.

Thus, viewing chemical knowledge from the perspective of the social system let us escape from a too narrow view that is blind for external influences and causes of progress of chemistry as a science. Conversely, the perspective of the material and the semiotic system of chemistry allows us to understand how social forms are shaped by the requirements and conditions of specific systems. Social forms were introduced in Georg Simmel's (1858–1918) classical text [262]. Social forms in Simmel's sense are

[33] A further semiotic concept Lavoisier popularised was that of conservation of matter. Although the idea was around from the beginning of the seventeenth century, it was Lavoisier who stated it for the first time and who showed its applications in chemistry [35].

Another important semiotic contribution was his notation to represent chemical reactions, which is essentially the one used today [35, 112]. For example, he wrote "must of grapes = carbonic acid + alcohol" [35] and added:

> We may consider the substances submitted to fermentation, and the products resulting from that operation, as forming an algebraic equation, and, by successively supposing each of the elements in this equation unknown, we can calculate their values in succession and thus verify our experiments by calculation and our calculation by experiment reciprocally [261].

abstract social relations that can link different individual content. Sociology is the analysis of such social forms. Important examples of social forms are collaboration, cooperation, coordination and competition. Clearly, they are fundamental forms for the social system of chemistry, as for those of the other sciences. Now, and this is our point here, their realisation is somewhat different between the various sciences. This results from the interaction of the social system with the corresponding material and semiotic systems.

In physics, we have gigantic projects like the Large Hadron Collider or other particle accelerators or the International Thermonuclear Experimental Reactor (ITER) project for nuclear fusion where hundreds, if not thousands of physicists and other scientists collaborate on a single task, like the detection of the Higgs boson. In biology, there was a coordinated international project, involving laboratories around the world, the genome sequencing project, and then also a competing, privately organised and financed such project. In the public version of the project, a clone-by-clone strategy was used, and scientists in different laboratories around the world were assigned different chromosomes, that is, parts of the human genome, for sequencing. The private project, initiated and led by Craig Venter (1946–), used a different approach, the shotgun method, where the entire genome was assembled from overlapping small fragments.[34] Both methods were successful, and so, the human genome was sequenced by two different approaches at the same time, in fact, even three years earlier than had been anticipated for the public project. Turning to theoretical physics, there is for instance the superstring community, to some extent informally coordinated by a leading figure like Edward Witten (1951–), that cooperates on the project of the unification of the known physical forces with a particular methodology. That methodology is different from that of competing approaches like loop quantum gravity. There is obviously scientific competition between the different approaches, but there also may emerge personal competition within a particular group.

In chemistry, these social forms are realised differently. In order to understand this more deeply, one should go beyond an outside ethnographic approach like that of [263] (which incidentally only looked at a physics facility (CERN) and a molecular biology laboratory) and understand this from the intrinsic requirements and structures of the material and the semiotic system of chemistry. Since the times of Liebig, there has been a particular type of successful organisation of a laboratory within which people collaborate or at

[34] The idea is to break up several copies of the genome into small pieces and to then reassemble it from the overlaps between the pieces from the different copies. For instance, if the sequence ACCGTAGGCCATGAACGTGG is once broken up into ACCGT and AGGCCATGAACGTGG and once into ACCGTAGGCCATG and AACGTGG, then the overlap AGGCCATG between one piece of the first and one of the second set suggests that the fragments should be combined into ACCGTAGGCCATGAACGTGG. Of course, since such short motifs can repeatedly occur in the genome, many more copies need to be fragmented in order to obtain a statistically reliable assembly of the genome from the fragments. The public project has dismissed that approach because it was believed that it could not achieve enough statistical significance.

least coordinate their use of the equipment.[35] Competition predominantly takes place between different such laboratories, sometimes fuelled by external distinctions like Nobel Prizes or other important awards and also by cooperation with the industry. An important arena of competition are grant applications, but since the public funding of science is differently organised in different countries, that competition has to correspondingly adapt to specific rules and regulations. But again, there are specific selection criteria for grant applications in chemistry that need not be shared by other fields. Large scale cooperations or collaborations do not seem to play a major role in chemistry, in contrast to most other scientific disciplines (elementary particle physics, astrophysics, molecular biology and neuroscience to name but a few of them), chemistry does not create large scale international projects in which thousands of scientists work together to systematically advance a field, like the human genome, proteome, connectome and other projects already discussed.

The interplay of the social, semiotic and material systems of chemistry have led chemists to engage in different projects mainly coordinated by research groups with much less people than those of the above mentioned projects in physics and biology. Interestingly, although chemistry lacks a rational research plan engaging and coordinating many research groups or a large fraction of the chemical community, there is a traversal historical trend in chemistry, namely the exponential expansion of the chemical space and its strong concentration on organic chemicals [4]. There is no written or agreed research agenda telling chemists to practice their science and to produce their substances the way they do it, but the social, semiotic and material conditions have led the discipline to follow this path. No single chemist thinks about contributing, with his/her extractions or synthesis, to an annual quota for the number of substances to meet the exponential report of new substances. Likewise, chemists do not care whether their compounds are based on carbon, hydrogen, nitrogen and oxygen. Nevertheless, over the history the community at large has expanded the chemical space in an exponential way, strongly favouring the production of substances based on those four organogenic elements [4].

[35] One could say that the synthesis and characterisation of new chemical elements is an instance of a large scale cooperation project in chemistry, and sometimes of competition. Nevertheless, projects of this sort are far from current mainstream chemistry (as discussed in Chap. 1), which is dedicated to the synthesis of complex molecular entities and of new materials. Moreover, the community of physicists, chemists and engineers supporting these projects is much smaller than that of physicists working in large scale projects such as those developed at the Large Hadron Collider.

4.6 Differences Among the Constitutive Systems of Chemical Knowledge and Their Temporalities

As in [19, 116], we see the importance of the interaction between these different systems, and a way to formalise this proceeds by looking at dynamical interactions between networks. These networks usually evolve, develop or change at different speeds. In general, in a dynamical system with variables evolving at different speed, in some phases the slow and in others the fast variables can dominate the dynamics.[36] In regular periods, the fast variables quickly relax to their equilibrium values, and the dynamics moves along the slow manifold. But when that slow manifold becomes unstable, the fast variables take over and let the system quickly jump to a different branch. Of course, for concrete applications, it is necessary to identify the relevant variables and model their dynamics to understand particular instabilities and quick or sudden transitions. Something similar may also apply to the interaction between networks with many variables, perhaps with different speeds both inside each network and between the networks, although more research is needed to make this point more precise.[37]

A general setting to discuss the differences among the systems compounding chemical knowledge and to address their temporalities is presented in Fig. 4.2. The social, semiotic and material systems of chemistry are based on objects and relations, which in the most general case can be modelled as networks (or hypernetworks, see Sect. 6.1). In this setting, the objects of each system are represented as nodes (or sets of them) in the network. Their relationships constitute the (hyper)edges of the network. In general, objects of the chemical knowledge may appear, merge, split, refine and disappear (Fig. 4.2).

Although the three systems that make chemical knowledge evolve, their formal structures and temporal dynamics are different. Networks of the social system (Fig. 3.1, see also Chap. 6) change over time, as not only new scientists enter, but also old ones retire or die and new industries and educational institutions show up and disappear. These social actors may split, merge and refine. The first is clear when, for instance, a researcher is the teacher of two or more researchers or when a company gives place to other companies, or even when some people of a company separate from it (Fig. 4.2). Merging occurs when two researchers are the teachers of a new researcher or when two or more companies become one, or when some persons decide to form a company, or even when a person joins a company. Refinements occur when a company changes its name and when a person changes its identity, for instance by changing its name when married or because of gender diversity reasons [266]. The temporality of the social system is associated to the duration of

[36] See for example the introduction to slow-fast dynamical systems in [264].

[37] As an application of network science to the history of science, it has been found [265] that innovations typically emerge in the periphery of a social scientific network and then perhaps spread to the centre and throughout the network. Here, the social network is constructed from bibliometric coauthorship data.

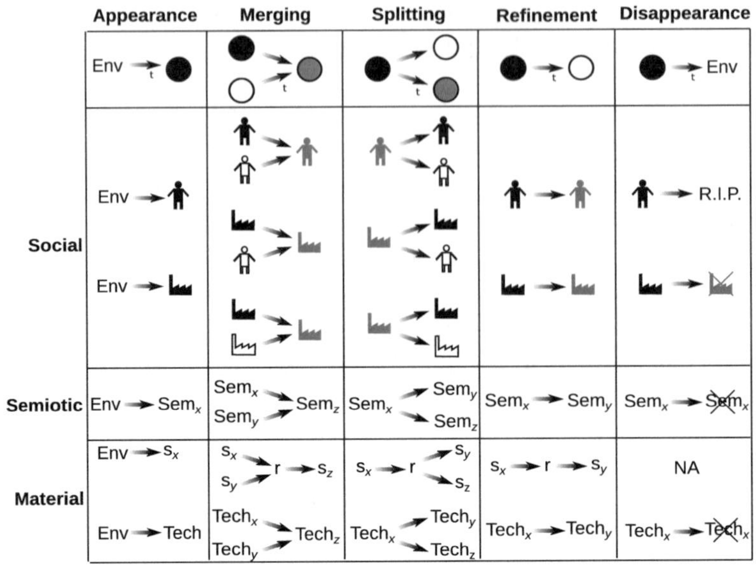

Fig. 4.2 General features of nodes in (hyper)networks used to model the constitutive systems of chemical knowledge. Arrows indicate temporal relationships between the initial event on the arrow tail and the final event at the arrow head. "Env" stands for the environment of chemistry (see Fig. 2.1). Circles represent objects of any of the three systems compounding chemical knowledge. "Sem$_i$" stands for the semiotic object i and "Tech$_i$" for the material-technologial object i. Likewise, "s$_i$" stands for substance i and r for a chemical reaction

social actors and their identity. If they are persons, its duration is driven by life expectancy, which has grown over time, leading to networks with people having more time to establish relationships with other actors. Governments, teaching and research institutions, as well as industries constitute another kind of social actors, whose durations may be long, as in the case of stable and lasting enterprises and forms of government and funding agencies, or short as in war-research institutes.[38]

As the semiotic system encompasses rather heterogeneous components and aspects, ranging from language tokens and grammars to concepts and theories, its formal modelling requires flexibility, for example by applying partially ordered structures, as well as algebraic and topological ones along with hypergraphs (Figs. 3.2, 3.3, and 3.4, see Chap. 6). Changes in the semiotic system sometimes are more or less complete replacements of established signs or research practices, as we have discussed in Chap. 3 (refinements); but they may include mergings and splittings. The periodic system is a case of merging, where information on chemical reactions and substances was integrated with abstract ideas of elements and with mathematical relations, namely order and similarity [39]. Cases of semiotic splitting abound in chemistry and often go hand-in-hand with disciplinary branching, from

[38] An interesting account of the history of the chemical industry is found in [267].

a presentist account, or which may be considered as mergings from a historical perspective. Examples include the rise of physical chemistry in the nineteenth century[39] from a whole body of mainstream chemistry, as seeing from a presentist account where current chemistry claims this territory.[40] Another example is the distinction between organic and inorganic chemistry, with its associated vitalistic and anthropocentric ideas going back to ancient times. Recent examples include biochemistry and material science, where novel conceptual structures arose. The presentist account indicates that biochemistry is part of chemistry, at least by regarding the semantics of the word "biochemistry"[41] and by considering the social structure supporting it, where often biochemistry departments are hosted by chemistry ones.[42] A recent case of semiotic refinement is the introduction of asymmetric organocatalysis, a research field initiated in 2000 [272]. The field of catalysis before this year was considered as devoted to the study and improvement of methods based upon a binary classification of catalysts. These substances not consumed in chemical reactions that increase the speed at which the chemical transformations occurs, were until 2000 considered as either complex and large molecules, such as enzymes, or as small transition metal-based compounds. Organocatalysis changed this assumption anchored in the pre-2000 semiotic framework of chemistry, by showing that small organic (non-metal-based) molecules can also assume relevant catalytic roles.[43] This constitutes a refinement of the fundamental concept of catalyst.

[39] Although the nineteenth century is associated with the beginning of physical chemistry, the term, and the very essence of this subdiscipline, was already evident in the eighteenth century through Mikhail Vasilyevich Lomonosov (1711–1765).

[40] Physics could also claim this area. In fact, there are two allied subdisciplines: physical chemistry and chemical physics. The former arose from studies on chemical thermodynamics, solutions (with and without electrolytes), chemical kinetics and related fields of research, typically sitting on an ontology of bulk substances, rather than on one about quasi-molecular and atomic species. Chemical physics, in turn, arose from a semiotic pressure of a growing community equipped with an ample spectrum of physical concepts that approached different chemical phenomena. This occurred in the first quarter of the twentieth century [268]. Although chemical physics research often sits on the quasi-molecular and atomic ontological level, the boundary of chemical physics and physical chemistry is not necessarily determined by this micro-macro distinction [269].

[41] As recently claimed by Stuart James Cantrill, Nature Chemistry editor [270].

[42] An interesting subject of research entails the social, semiotic and material conditions leading to the rise of biochemistry, which have involved tensions within chemistry, as evident, for instance in [271].

[43] The term "asymmetric organocatalysis" indicates that some organic catalysts lead to products where an enantiomer is favoured over the other. That is, breaking the symmetry of obtaining reactions products in similar proportions.

How semiotic objects appear in the system is one of the subjects of semiotic studies in chemistry, but the appearance of these objects is definitely driven by the experiences upon matter transformation (see Chap. 1). Semiotic objects also disappear, as in the case of theories or concepts no longer used [273, 274].[44]

Semiotic changes have different temporalities, which are often associated to the social system in which the change occurs and is finally accepted. Sociologists of science study these times of appropriation, for instance through the cases of sleeping beauties (see footnote 30 in Chap. 1). These temporalities may be relatively short as in the case of the replacement of the phlogiston theory around 1785 thanks to Lavoisier's works of 1783 and preceding years [276] (p. 152). Semiotic temporalities may also be much larger. In Chap. 1 we provided several examples of those cases, where we presented an extreme case of a semiotic change requiring about a century to be accepted.

The material system can be modelled as a directed hypergraph of substances, reactions and associated technologies that grows in time (Figs. 1.2 and 3.5), where neither substances nor reactions have disappeared during the last two centuries, as chemical knowledge is preserved in repositories and transmitted across generations through established teaching and research practices operating in social networks. Technologies follow a different path as they often disappear, for instance chemical tests used for molecular structural characterisations, which were replaced in the 1950s by spectroscopic technologies [155, 156].[45] Technologies also undergo refinements when they are transformed into new ones by improvements. This is the case of laboratory glassware, used in the eighteenth century and widely popularised in the nineteenth century thanks to the political and economic conditions allowing the transport of Bohemian potash glass to different European centres of chemical activity [278].

Objects of the material system appear, for instance by extractions or by the introduction of technologies, which both often come from the environments of chemical knowledge (see Chap. 1). The material system experiments mergings, splittings and refinements, as well. They are actually at the core of the expansion of the chemical space when the material objects are substances. This is so because of the very definition of chemical reaction that entails the temporal combinations and transformations of substances into

[44] Establishing the boundary between refinements and disappearances is difficult and it has motivated research and strong debates in philosophy of science [143, 275]. We mention the abandonment of phlogiston theory as a case of semiotic disappearance, but we do not mean it disappeared as a historical object, rather as a theory guiding the evolution of chemical knowledge, especially after 1783, thanks to Lavoisier's experiments and interpretations leading to its eradication from the semiotic system. Interestingly, Chang has questioned the sudden nature of this transition.

[45] An instance of these tests is the iodoform test, used to determine whether a compound contains the aceto group (CH_3CO–) joined to either carbon or hydrogen. It also indicates whether the compound is oxidised under the conditions of the reaction to derivatives containing the aceto group [277].

new ones.[46] Mergings are addition reactions, while splittings are decomposition ones. Refinements, in turn, occur with rearrangement reactions.[47] Nevertheless, as the material system involves more than substances and reactions, for instance technologies, mergings imply devising new apparatus or laboratory techniques as well, where different material pieces are put together with a given purpose. Still in the field of technologies, splittings may be considered as derived technologies and instrumental methods, for instance the rise of gas and liquid chromatography from either column or paper chromatography.[48]

The temporality of the material system is mainly driven by the expansion of the chemical space (discussed in Chap. 1), which unfolds at a very stable historical rate. The material system also evolves according to the speed of generation and adoption of new technologies. Interestingly, chemistry is one of the sciences providing more technologies (see Chap. 1) but one of the sciences that takes the longest to accept them.[49] An interesting research subject is measuring the seed of chemical technological growth to have a more complete picture of the temporality and expansion speed of the material system of chemical knowledge.

An instance of the different speeds of the three constitutive systems of chemical knowledge, even for the same historical periods, is the expansion of the material system in the sixteenth century and the stagnation of the semiotic system, as evidenced in the relatively minor theoretical advances of that century [35].

Likewise, the coupling of the systems of chemistry to external dynamics is different. For instance, the fast dynamical material system is coupled to a slow social system limited by the growth of the chemical community. In turn this community is associated with other social constraints such as how appealing practising or studying chemistry has been over the course of history [109].[50]

---

[46] See Sect. 1.2.

[47] A hypothesis to be tested by computational analyses of the chemical space is whether the analytical period of chemistry can be characterised by a large proportion of splittings, whereas the synthetic one by more mergings.

[48] The origin of this separation technique is still a matter of debate. Some argue that it was paper chromatography [279] instead of the widespread idea that it started with column chromatography [280].

[49] As sleeping beauties not only refer to semiotic changes but also to technological advances (see [70] and discussion in Chap. 1), they can be taken as a proxy for technological appropriation.

[50] Schummer discusses the difficulties of estimating the size of the active chemical community [109]. In his approach, an active chemist is one providing new chemicals through scientific publications.

4.7 Environments of the Constitutive Systems of Chemical Knowledge

The development of chemistry is coupled to that of other sciences. Atomism was played back and forth between chemistry and physics [91].[51] And while originally biology had an important impact on chemistry as the source of many substances, in the twentieth century, chemistry in turn dramatically changed biological research. In general, as a complex system, chemical knowledge depends on regularities of the environment for maintaining its internal structures and processes. A straightforward part of the environment is given by the relation of the social system of chemical knowledge with other kinds of knowledge, for instance physical or biological (Fig. 2.1). Thus, a social actor may take part in a social system producing chemical knowledge as well as physical and/or biological knowledge. One may think of the National Science Foundation or the *Deutsche Forschungsgemeinschaft* supporting current chemical, physical and biological projects, as well as monarchies and other forms of sociopolitical organisation that have supported chemistry studies. In fact, such agencies or organisations influence and sometimes decisively shape both the social and the material system, as they support both positions for scientists and finance laboratory equipment. Political and economic decisions indicate which research directions should be given priority and consequently be heavily funded, at the expense of others. They usually, however, do not directly affect the semiotic system. But the influence on the semiotic system can be indirect, when new research foci like material science lead to new types of data organisation in data storage facilities. Actually, nowadays, data organisation often constitutes a research topic of its own, and then funding agencies can even directly influence the semiotic system.

Another part of the environment of the chemical knowledge could be the semiotic systems of mathematics, physics and physiology, where an interchange of signs and practices has taken place [42, 71, 282]. Likewise, physiology and biology have been recurrent sources of new substances for the material system, and currently chemistry, in a very prolific partnership with physics, is providing physics and engineering with materials of unprecedented properties. How regular these contacts and interchanges with the environment have been is a question to resolve, as well as to assess its impact on the evolution of chemical knowledge. The relations of the system of chemical knowledge with the environment are represented with dotted lines in Fig. 2.1.

A further part of the environment of the system of chemical knowledge is its interaction with the social system at large, which includes the social system of chemistry. This relation is evident, for example, in the contributions of chemistry with its drug discovery output to steadily increasing life expectancy [31]. This is a symmetric relation, as social pressure leads to producing medicines with the associated effect of emphasising the exploration

[51] In [91, 281] Rocke shows how there were two parallel approaches to atomism reaching maturity in the nineteenth century: chemical and physical atomism, which influenced each other.

of a particular region of the chemical space [67, 240].[52] Likewise, social demands led chemists to synthesise fluorochlorocarbons as refrigerants with their devastating effects on the ozone layer leading to environmental and health problems, which in turn motivated further chemical research for alternative refrigerants.[53] Another instance of this interaction is evident through the social relevance of the Nobel prize, which influences research funding and the scientific interests of new generations [283]. The relationship between the semiotic and the social system is evident through the incorporation of chemical symbols and practices in daily life. For example the iconic image of a scientist, actually a mad-scientist, is that of a chemist with his/her white-coat and a test tube or a flask in his/her hands [230].[54] The periodic table is another case, which is regarded as an icon of science and the most celebrated icon of chemistry [39]. The social system of chemistry has also influenced the semiotics of the discipline, for instance through the biased negative image of chemistry as a science producing pollutants and weapons, which has led to even drop the word "chemistry" from new chemical specialities such as new materials, nanotechnology and others [128].

Besides the importance of these studies to understand some particularities of the separated role of each one of the three systems upon knowledge production, there is no report of a combined approach, including the simultaneous action of the three systems, to understand and model the evolution of chemical knowledge.[55] This is the programme we put forward in this book. Obviously, there are many instances where a closer study is possible and will be insightful. For example, around 1800, many chemists in Germany had been trained and often were still working as pharmacists. Kant's colleague, the professor of medicine Karl Gottfried Hagen was such a pharmacist, and he founded the

52 This pressure is particularly evident in the COVID-19 pandemic, where funding agencies have adjusted their priorities to the search for vaccines and related medicines, as well as to explore social, economic and psychological effects of this event.

53 Interesting discussions on the interaction between the social and the material system of chemical knowledge are found in [128] and in the 2006 HYLE issues devoted to the subject.

54 It has been shown how this icon was associated to alchemy and then to chemistry as a sign of charlatanry [230]. The icon was indeed borrowed from seventeenth-century physicians. At that time, it began to become clear that better diagnosis practices should go beyond urology and those still relying exclusively on urology practices were considered charlatans. The iconic figure of a man holding a flask and observing its urine content was then that of an old-fashioned physician, a charlatan. This image was then assigned to alchemists, which also held a bad reputation for their practices. Interestingly, popularisers of this semiotic transference were dominant figures of the seventeenth-century social media, namely painters. The association of alchemy and charlatanry goes as far back as 292 A.D., when for instance the Emperor Diocletian ordered to burn all alchemical books as a consequence of the falsification of gold and the issues this caused for the Roman economy [35].

55 There are also few studies on the interactions of chemistry with industry, politics, and society [2]. One of these is a recent analysis and quantification of the role of World Wars upon chemical production [4].

first pharmaceutical-chemical laboratory in Germany, at the University of Königsberg in Prussia. The discoverer of morphine, Friedrich Wilhelm Sertürner (1783–1841), ran pharmacies (and had to struggle with the resulting economic problems) throughout his life, see [284]. In fact, Sertürner had isolated the first alkaloid, and he called it *morphium* after Morpheus, the Greek god of dreams. It should then be studied how such representatives of the social system, which often had never had the chance to attend a university lecture, with their often profound practical knowledge and their way of thinking shaped the semiotic system of chemistry and how the types of substances they investigated and their research practices provided a direction for the material system, and how in turn this contributed to the professionalisation of the profession of pharmacists.

Similar questions can be asked about the relation between chemical science and chemical industry. This is of course a well studied subject, but we are asking for investigations that analyse the relations and interactions between the three different systems.

In Part II, we present the kinds and sources of data related to this programme and the suitable formalisms to model the evolution of chemical knowledge.

Part II

Data and Mathematical and Computational Formalisms for the Analysis of Chemical Knowledge

Data

5

Abstract

We present the different sources of digitised data to carry out computational studies on the evolution of chemical knowledge. We also provide information on important data requiring digitisation to be incorporated as part of the corpus for studies on the evolution of chemical knowledge. For each of the semiotic, social and material systems of chemistry, these data sources are discussed.

Large amounts of highly reliable historical data are key to studying the evolution of chemical knowledge. Studies addressing particular issues of this evolution are not abundant mainly because of the lack of publicly available data. However, in the present computational era large amounts of non-born digital data such as printed historical records have been digitised, which in combination with born digital data of the present times constitute an excellent corpus for analysing the evolution of chemical knowledge. In the following sections we describe some of the available sources of information for each one of the systems of chemical knowledge and in Chap. 6 we present the formal methods to operate upon these data to solve the different questions posed in previous chapters.

J. Jost, G. Restrepo, *The Evolution of Chemical Knowledge*, Wissenschaft und Philosophie – Science and Philosophy – Sciences et Philosophie,
https://doi.org/10.1007/978-3-031-10094-9_5

5.1 Material System

Likely the best documented and standardised system, although not publicly available, is the material system, which began to be recorded, curated and annotated as early as the thirteenth century.[1] Instances of unique quality and scope, high importance and wide influence are the handbooks edited by Gmelin and Beilstein. Gmelin published the first edition of his *Handbuch der theoretischen Chemie* in 1817. Later in that century, it was found that, because of its spectacular growth, organic chemistry needed a handbook of its own, and Beilstein then published the first edition of the *Handbuch der organischen Chemie* in 1881/1882.[2] The handbook started by Gmelin then changed its name to *Handbuch der anorganischen Chemie*, and new editions of both handbooks then appeared regularly to document the progress of chemistry. These handbooks had a decisive influence on the evolution of chemistry and its unity as a field. They were complemented between 1830 and 1969 with the *Chemisches Zentralblatt*. Between 1922 and 1997, the Gmelin Handbook was edited and later also a corresponding online data base was operated by the Gmelin-Institut für Anorganische Chemie und Grenzgebiete which became an institute of the Max Planck Society in 1948 and which employed up to 120 people. When the institute was closed in 1997, 760 volumes with 240,000 printed pages of the handbook had appeared. Since 1981, the handbook was published exclusively in English, and in 1990, its name was changed to *Gmelin Handbook of Inorganic and Organometallic Chemistry*. The Gmelin and Beilstein handbooks were digitised and merged in what is called today Reaxys©, a large repository of chemical information that also includes weekly updates from 16,400 journals and patents. The database is owned by Elsevier. Another source of information is SciFinder, owned by the American Chemical Society, with daily updates. This latter includes the *Chemisches Zentralblatt* in digitised form and translated from German into English. Most of the original German issues of the Zentralblatt are publicly available in electronic form in the Internet Archive [286] and in several libraries linked to the German Wikipedia site of the publication [287].

Our particular experience analysing large amounts of chemical information is with Reaxys©, which for the material system provides information about substances and their properties; likewise about primary sources where these properties and the reactions reporting the use or production of the involved substances are published. Information about reactions, along with reaction conditions is also provided, that is temperatures, catalysts, pressures, solvents and reaction times, among others.

Sources of information on failed reactions, besides the papers reporting them in their supplementary material, are The All Results Journals: Chemistry and the blogs and wikis of chemists associated to the open-notebook science movement. Richer sources of failed reactions are printed notebooks of chemical laboratories and graduate theses,

[1] See Chap. 4 for more details.

[2] For information about Beilstein see [285].

which need to be digitised. Moreover, chemical industry [288] electronic lab notebooks [289] constitute another source of information. Thus, although the information on failed reactions is out there, digitisation technologies need to be applied to the available printed sources, which require developing standards and protocols for the annotation process [290–293].[3]

An interesting data movement is the national research data infrastructure initiative for chemistry in Germany (NFDI4Chem) [293–296], which aims at "creating a general research data management portal and hereby connecting already existing infrastructure as well as to foster the cultural change in chemistry towards digitalization by developing general minimum information standards for all methods used and teaching RDM [Research Data Management] principles to the community" [293]. The challenge consists in integrating the semiotic data discussed before into this infrastructure, as well as the social data we shall discuss in the next section.

A further source of information for the material system is Ewing's Analytical Instrumentation Handbook [297], which in its several editions constitute a reliable source of information for chemical instrumentation and its evolution. Likewise, modern subdisciplines such as material sciences and high pressure chemistry possess well documented literature on their instrumentation. For instance the Handbook of Instrumentation and Techniques for Semiconductor Nanostructure Characterization [298] for the first and the Handbook of Techniques in High-Pressure Research and Engineering [299] for the second. Reaxys is again a suitable source of information on chemical instrumentation, as it annotates reactions reported over the history with the conditions and equipment used for conducting them and for separating and purifying their products.

5.2 Social System

Databases such as Reaxys© play also an important role for the social system of chemistry. Through primary sources, Reaxys© associates authors and institutions to substances, their properties, reactions and reaction conditions. The bibliographic information allows gathering information on publication channels (journals or patents) used to spread new chemical knowledge. We have analysed, for example, the rise and fall of certain journals in the publication of new chemical substances since 1800 until today [4]. Other sources of chemical bibliographic information are the Web of Science [300] and Dimensions [301, 302], which are databases conveying information about publications, authors,

[3] These standards require considering the different facets of failed reactions. That is, experimental settings leading to no chemical change at all, or to those actually leading to new substances, but not to the expected ones. This second case is often considered as a failed reaction as the theoretical setting in which the reaction is thought of and conducted does not lead to the theoretically expected product. Therefore, the social and semiotic dimensions of failed reactions need to be considered when devising standards and protocols for their annotation.

disciplines and subdisciplines, citations, references and affiliations, to list but a few sorts of data.

An important complement to the bibliographic approach to estimate the size of the chemical community over time requires the automatisation of traditional methods used by historians. This entails selecting and prioritising possible sources of information such as membership to chemical societies, at least in Europe and North America and currently in Asia. Important sources of information in this direction are, for instance for the European chemical community, the book [147]. For the American chemical community, sources such as the Collective Author Indexes of the American Chemical Society are instrumental. The social system of chemistry also requires gathering information about registered students for chemistry studies, where repositories of universities are central.

A central issue in the annotation of the social system is posed by homonyms, which becomes more dramatic given the growing number of Chinese authors appearing in the current chemical literature.[4] The problem is compounded by the fact that the Pinyin romanisation of Chinese characters can lead to the same transcript for phonetically similar names that are represented by different Chinese characters. Also, foreign names get sinised in China. Thus, for instance, the Jesuit scholar Matteo Ricci (1552–1610) was named Li Matou, that is, obtained the Chinese surname Li. Thus, seemingly many papers are authored by the same authors when in fact they are contributed by different authors. An instance of the magnitude of this effect is seen, for instance, in the many papers assigned to any combination of the most popular Chinese surname with some initials. Examples include Wang Y. with 141,333 publications in the ISI Web of Science (26th October 2021) between 2000 and 2021 and Wang J. with 115,639 for the same period. This contrast with a very popular English surname such as Smith, with only 11,040 for Smith J. and 10,647 for Smith A. for the same period. This documentation problem is a matter of current research.[5] Some of the computational approaches to solve it use, for instance, machine learning methods [304].

Further data sources for the social system will require a combination of historical methods with current approaches to computationally characterise an author's identity, for instance through ORCIDs (Open Research and Contributor ID) [305]. This approach will assign a unique ORCID to past authors such as Glauber, Lavoisier and Kekulé, which will allow for conducting computational studies on the influence of these and other chemists, as well as in the spread of their findings.

---

[4] The distribution of Chinese surnames is strongly concentrated on very few items in comparison with the rest of the world. The number of surnames in China is around 6000 and about 90% of the population share just 100 of those surnames; while for instance in 2000 the US census reported more than 6.2 million surnames, the majority of them reported only once [303]. Wang, Li, Zhang, Liu and Chen are the five most common surnames in China, shared by more than 30% of the population [303]. This has cultural and historical reasons, among them the ancestral rites. Therefore, Chinese surnames are not created by assigning new characters as it may occur for Western surnames [303].

[5] Funnily, a naive search produces et al. as the most prolific coauthor.

Suitable sources of information for the role of the chemical industry in the production of new chemicals and its relationship with the academic world are the archives of chemical companies, for example of Bayer-Leverkusen and BASF-Ludwigshafen, to pick out some pivotal companies working in close collaboration with academics in the nineteenth century. The organisation of the information related to the chemical industry must consider the mergings and splittings of companies, which is well documented in the weekly issues of Chemical & Engineering News and spans the salient aspects of the chemical industry since 1923. For the period before, especially the nineteenth century, the *Zeitschrift für Angewandte Chemie* is a suitable source of information. The successful gathering of this information, its curation and annotation require the expertise and methods of specialised historians, which need to be combined with suitable optical character recognition methods for the digitisation of printed archives.

Another source of information is the International Institute of Social History [306], whose databases offer, for example, information on thousands of occupational titles from countries and languages around the world from the sixteenth to the twentieth century. Clio Infra [307] also holds interconnected databases containing worldwide data on social, economic, and institutional indicators for the past five centuries, with special attention to the past 200 years. In particular, excellent data sources for historical prices of starting materials are the corporate archives and the catalogues of fine chemicals such as those of Aldrich, for instance, for non-born digital information of our current times.

5.3 Semiotic System

Although there are some studies on the semiotics of chemistry for some particular periods [31, 71, 118, 138], there is no suitable and complete database for the evolution of semiotics in chemistry. Digital versions of leading journals of the past plus the *Chemisches Zentralblatt* are important sources of these data, as they gather the nomenclature, signs and relations of those objects used by chemists over history.

A first approach to create the semiotic database of chemistry therefore requires the digitisation of images of apparatus and laboratory equipment, as well as of tables, diagrams, reaction schemes and molecular structures among other semiotic objects of chemistry. This digitisation must go beyond the storage of those semiotic objects and must incorporate their intention and affordance. For example, why a molecular structure is sometimes drawn highlighting some structural aspects, whereas in others the structure is rotated to stress some other aspects? Often these representations obey an intention of the author. In other cases they are driven by the available printing technology. In some other cases the representation follows from aesthetic stances. This is the intention that needs to be incorporated in the semiotic database.[6] It can be gauged by encoding the semiotic object along with its context, which is often provided by the text accompanying the object. In this case, for instance, natural language processing tools such as sentiment analysis turn instrumental for these purposes.

Reaxys© and SciFinder are also important sources of historical common names for substances. At any rate, building up a semiotic database is a must, where close collaboration with historians of chemistry and linguists is required, whose methods are needed to extract data and to build up temporal networks of concepts, for example. Current sources of born-digital chemical semiotic information [308] are the blogs and wikis of the open-notebook science movement, which is not only devoted to the synthetic activity in chemistry but to other areas such as analytical and theoretical chemistry [290, 309]. Even if these data are already in electronic form, the creation of dynamical databases requires protocols and normalisation to turn the information into an actual source for semiotic studies in chemistry.

It is a general question how the modern tools of automatic data analysis and data storage also change the semiotic system of chemistry. In the end, the main purpose of chemical nomenclature might no longer be the compact codification of chemically relevant information for humans, but the ease of its processing and storage by computer systems and perhaps its interpretability by robots. In any case, a useful semiotic system will always require universal conventions and standards.

[6] Currently, chemoinformatics is experiencing a transition from molecules to reactions, where molecules are not any longer the only focus for codification and computing but also reactions, their classification, encoding and processing. An instance of this move is the broadening of the InChI codes to RInChI codes [50], designed to encode information on chemical reactions at different levels. RInChI does not incorporate intention so far, but the RInChI structure allows for further additions of intention, which may link the chemoinformatics notation for reactions with those suitable for computational historical studies.

Methods

6

Abstract

We present the mathematical and computational frameworks to analyse the data pertaining to the material, social and semiotic systems of chemistry. These include, but are not limited to, hypergraphs, complexity measures, time series analysis, statistical physics methods, agent-based modelling, language theory, formal concept analysis, machine learning and text analysis and computational semiotics. We outline possible uses of these formal tools. This chapter is therefore partly speculative, if not visionary.

The introduction of this research programme to analyse the evolution of chemical knowledge has been presented through a combination of historical ideas with a computational and formal background. In this chapter we provide further details about the methodologies and theoretical settings supporting our formal stance to treat history of chemistry information. The varied and heterogeneous nature of the formalisms discussed in this chapter constitute suitable tools to analyse the evolution of chemical knowledge. Nonetheless, nothing restricts the application of other theories or methods, let alone the development of new ones to treat the particularities of chemical knowledge.[1]

[1] Some examples of mathematical methods and theories developed to solve chemical questions are discussed in [310]. For some current mathematical theories of application in chemistry see [282, 311] and specialised journals such as MATCH Communications in Mathematical and in Computer Chemistry and the Journal of Mathematical Chemistry.

J. Jost, G. Restrepo, *The Evolution of Chemical Knowledge*, Wissenschaft und Philosophie – Science and Philosophy – Sciences et Philosophie,
https://doi.org/10.1007/978-3-031-10094-9_6

6.1 Hypergraphs

Network analysis is one of the best developed fields in complex systems theory [312]. Traditionally, networks are modelled as graphs, and therefore include only relationships between pairs of elements. Currently, an extension is under intensive development: hypergraph models. These structures model relationships among sets, which may or not contain internal structure. We model the evolution of chemical knowledge as a dynamical hypergraph [313] made of the simultaneous relationships [314] among the three systems (structured sets) of chemistry (Fig. 2.1). Hypergraphs [315] or, in general, hyper networks [316, 317], also model each one of the compounding systems of chemical knowledge. In a hyper network the objects that are related are not restricted to only binary relations. A compelling chemical example is the use of hypergraphs to model reactions (Fig. 1.2) [318, 319]. When modelled through graphs, the notion of starting materials fades away as well as the notion of products, as these important set-theoretical notions are disregarded [320, 321]. In contrast, the hypergraph setting considers relations among subsets, that is, the set of substrates leads to the subset of products while keeping the identity of the individual substances.[2] A *directed* hyperedge then links the substrates with the products and in this way represents the corresponding reaction. The hypergraph relating the set of substrates with that of products can also be regarded as a traditional directed graph, but this requires the introduction of a new vertex in the graph for each reaction.[3] In that representation, each individual substrate of a reaction leads to the reaction and the reaction leads to each particular product. Importantly, mathematical concepts should be developed for *directed* hypergraphs [318, 319, 322], as chemical reactions have their natural direction from substrates to products.

Another instance of hypergraphs in chemistry is the recent formalisation of the periodic system of chemical elements as an ordered hypergraph, where similar elements are grouped in subsets, which are ordered according to the order relationships of the chemical elements [323]. Likewise, hypergraphs can also be applied to define challenging concepts such as aromaticity. In this setting benzene, for instance, is defined as a set of six equivalent carbon-hydrogen bonds, where the six carbon atoms form an equivalence class of atoms.

Hypergraphs can also model the social system of chemical knowledge, where chemists may belong to universities or to industries, for instance, or both, but they need to retain their identity for further analyses. Institutions are sets of individuals in this setting. Likewise, every co-authored paper defines a set of authors and a collaboration network

---

[2] Although there is still a binary relation, that one between substrates and products, the main difference with the traditional approach of network theory is that substrates constitute a set and products another one. In Fig. 1.2 these related sets are {C,D} → {A,B}, {B,E} → {F} and {F,G}→ {D,H}.

[3] For that purpose, one constructs a bipartite graph, with one class of vertices corresponding to the substances, and the other to the reactions. A substance vertex is connected to a reaction vertex when the substance participates in the reaction.

therefore depicts relationships among individuals while providing relevant information on the sizes of the publications (number of authors).[4] At the semiotic level, hypergraphs also turn central, as, for example, only binary relationships among semiotic objects such as words are unable to provide meaning.

To analyse the evolution of hypergraphs, suitable statistics describing their dynamics come into play. Only through these statistics quantitative statements about hypergraph evolution can be gauged as a consequence of the enormous size of the networks associated to the chemical knowledge.[5] Statistics for traditional networks abound, for instance degree distributions, betweenness centrality, spectra of the graph Laplacian, curvatures and several others [325]. Although there are some for hypergraphs [326], such as vertex and hyperedge degrees, clustering coefficients, spectral properties [315] and, more recently, curvatures [318, 322, 327, 328], further statistics have to be developed to attain a complete characterisation of the evolution of chemical knowledge.

The theory of dynamical networks, which results from the interplay between (hyper)graph theory and nonlinear dynamics [325], becomes instrumental for these studies. In this setting, objects are endowed with time-dependent quantities which change and evolve according to dynamical rules, which involve not only self-interactions, but also interactions with neighbouring objects. The aim is to understand the global dynamics of the modelled system, which emerges from the coupled system of dynamical equations resulting from the interactions. Dynamical networks also consider how evolution rules shape the dynamics of a network by allowing for dynamical network topologies. This

[4] As in many disciplines, including chemistry, the order of the authors of a paper contains information about social hierarchies, a corresponding refinement of the notion of a hypergraph should provide additional insight. Such a suitable refinement may be provided by ordered hypergraphs, as discussed in [323].

[5] For example, the material system accounts for more than 57 million chemical reactions and more than 170 million substances (Reaxys© 2022, January) [3]. See footnote 2 in Chap. 1. There are more than 6 million chemistry papers published over the history of chemistry (Fig. 1.1), and the number of semiotic signs must be much larger than 170 million as there have been periods in history where the same substance was represented by different signs. For example, Brock (p. 253) [29] and Bensaude-Vincent [324] remark that Kekulé referred to 19 different formulae for acetic acid in his *Lehrbuch der organischen Chemie*, published in 1859. In [92] we show how current chemical data, which often normalises those chemical formulae according to the standard set of atomic weights, can be adjusted to recover the different formulae in vogue before 1860, which resulted from different competing sets of atomic weights. The year 1860 is central as after the Karlsruhe conference at that year, a stable set of atomic weights was adopted. The unique and canonical representation of a substance is still a shortcoming, and it is in fact the motivation behind the InChIs (see footnote 33 in Chap. 1). For instance, the same molecular structure may be represented and stored by different connection tables or SMILES (see footnote 13 in Chap. 3), which require canonicalisation algorithms to reduce their numbers. If to these semiotic objects we add those of theories and concepts and their high order relations, the enormous size of the semiotic network becomes clear. The chemical community which with its institutions at a given time forms the social network of chemistry is smaller, but still very large in absolute terms.

formalism allows for studying couplings of interactions running on different time scales, where object dynamics are fast and the dynamics of the interaction parameters and the global emerging dynamic are slow. Several examples of different temporalities of interactions in the system of chemical knowledge were discussed in Chap. 4. Dynamical networks can be also used to explore the topology of the social system of chemistry and to assess to which extent such a topology permits the presence of gatekeepers [146] for the rise of disciplines and the spread of ideas. Such tools also enable us to better understand the emergence and diffusion of innovations. In fact, innovations usually originate not in the centre, but rather at the periphery of scientific networks [265].

While usually, ordinary graphs are taken as underlying dynamical networks, recently also dynamical networks on hypergraphs are being studied and reveal surprising phenomena, for instance those discussed in [313].

Besides statistics characterising static stages of hypergraphs in their evolutive paths, generative models for the dynamics of these structures are required. They involve, necessarily, random models to contrast with and more sophisticated ones involving particular features of each system to attain accurate dynamical models. Typical examples of these models for traditional networks include the Erdős–Rényi, Barabási–Albert and small world settings, among others [312]. These generative models are required to quantify how different the chemical knowledge and its systems are from a random evolution and to pinpoint the simplest model able to match the data on chemical knowledge.

Generative models of hypergraphs may be used to solve questions such as the underlying mechanism driving the exponential growth of the chemical space. Simple models may be based on the historical frequency of participation of substances in reactions and on the frequency of appearance of new reactions, which may shed light on the future of the network of chemical reactions, the size and geometry of the chemical space. These generative models could be further refined by introducing molecular structural aspects leading to the different frequencies of participation of substances in reactions and of appearance of new reactions.[6]

The same generative models may turn instrumental for understanding how concepts are intertwined giving place to new knowledge in networks of concepts. Was the structure of this conceptual network different in the early years of chemistry, when compared with

[6] This remits to think in models such as the Barabási–Albert one [329], which in this setting means that the likelihood of participation of a substance in a reaction is proportional to its historical participation in other reactions. In fact, we found in [4] that a large part of the chemical space has been expanded by combining well-known chemicals with new substances. But perhaps chemistry requires further additional conditions, for instance by including the functional groups present in the molecule. This additional condition would rule out cases of reactions among chemically equivalent substances, say, between two alcohols, and would prefer reactions among chemically dissimilar substances, say alcohols and carboxylic acids. Similar arguments may be used to model the appearance of new reactions, this time based on the participation and functional groups of the substances taking part in the transformation.

today's network? Or is rather the conceptual network a topological invariant of chemical knowledge? To which extent is this network affected by the appearance of new concepts, by the merging and splitting of some of them? Can we model the appearance and fading of concepts in chemistry based on the historical records of this network? Further questions based on the hypergraph modelling of the semiotic system of chemistry involve the effects of the environments of chemical knowledge such as knowledge from physics, mathematics and biology. To which extent does such knowledge affect the conceptual network of chemistry?

A social question generative models may help to solve is whether there have been historical periods where social forms of organisation in chemistry have been more regular or rather close to be random. Solving this question invites to think about the interplay of the social organisation with the material and semiotic system, which is a central question for the evolution of chemical knowledge.

6.2 Complexity Measures

We have highlighted the importance of complex systems in the framework of chemical knowledge, which requires defining complexity in chemistry and measuring it for the chemical knowledge and for its subsystems. Theoretical approaches to complexity address the following issues: (i) How difficult it is to generate or to describe a system/structure, (ii) how difficult it is to understand a system/structure and (iii) to what extent a system is more than the sum of its parts [330]. These approaches require selecting the relevant variables characterising chemical knowledge as a whole and selecting those describing each one of the three systems. This entails building up large repositories of historical data, as discussed in Chap. 5. The approaches to complexity also require modelling the growth and evolution of chemical knowledge and of its systems.[7] Therefore selecting the relevant statistical features of these models is crucial. Once these statistics are selected, they can be incorporated into the general settings for measuring complexity [120,330,332]. Following (iii), a set S of systems to which complexity is to be measured and a set of parts D are assumed. Hence, given a system, the process of extracting its parts is a mapping from S to D. As the aim is contrasting whether the system is more than the sum of its parts, another mapping is considered, in the reverse direction from D to S. The complexity measure for the system of interest then compares the system resulting from the composition of these two mappings with the original one. If the two systems coincide, then the whole equals the sum of its parts and the system is said to display no complexity. Complexity is quantified through deviation from that state [330]. In this setting, each of the systems of chemistry, including the whole system of chemistry, can be assigned a complexity

[7] As Epstein and Axtell argue [331], we could approach scientific explanation through the growth of the system under study.

value based on their parts.[8] This depends on the chosen level of resolution. For instance, the complexity of the material system can be addressed by selecting bulk substances and their reactions as its parts, or at the quasi-molecular level, where atomic ensembles become the parts. Questions that arise here involve contrasting these complexities. Are descriptions of the material system at the quasi-molecular level more complex than those based on the ontology of bulk substances? By introducing spectroscopic methods and even computational technology, how much were these complexities affected? Measuring complexity at the social system entails also different levels. Which system develops more complexity; one based on institutions, or countries, or a system described at the interaction level of people? Similar questions arise in the semiotic system, with levels of description including theories, concepts and fundamental signs. All in all, complexity measures help to understand the historical unfolding of the complexity of chemical knowledge and its constitutive systems. These measures turn instrumental to analyse periods of normal science and highly innovative ones. Do changes in complexity patterns of chemical knowledge correspond to breakthroughs in one of its constitutive systems or are they contemporary with changes in the way of practising chemistry as a whole?

Also, complexity measures of the type discussed here bound the robustness of a system [333], defined as the ability to compensate for the loss or malfunction of important parts. We may then analyse situations of failure within one or several of our three systems in terms of the complexity, or perhaps lack of complexity, that the system(s) had built up. Its steady growth rate might indicate that the chemical system as a whole is remarkably robust, and by implication therefore also highly complex.

6.3 Time Series Analysis

A key mathematical tool to analyse dynamical data is time series analysis [334, 335], which given a stochastic temporal function makes it possible to forecast and study trends [336], changes and evolving behaviours, for instance [337], as well as to assess associations between temporal distributions of different provenance [337]. A key feature of time series analysis is its ability to quantify the systematic features of temporal data by distinguishing them from random data variation. This means that it provides a mathematical tool to differentiate general historical trends from contingent events. This technique has been recently recognised by historians of science as a suitable method to assist their narratives and to find causal relationships [338, 339]. A recent example of application of time series analysis to the evolution of the material system of chemical knowledge was the quantification of the historical growth rate of the annual output of

[8] More particular examples include measuring the complexity of chemical reactions, where not only substances and reaction conditions are involved as parts of the system, but laboratories and their available technologies.

new substances [4]. It was found that the process of reporting new substances follows a heteroskedasticity model, that is, made of periods of different variability (Fig. 1.3), which we dubbed regimes. The first regime (1800–1860) presented the maximal variability in the annual number of new substances reported, which is the result of the small size of the community and the incipient infrastructure to regularise chemical production. The second regime (1860–1980) presented a reduction of this variability, which went hand-in-hand with the institutionalisation of chemistry, and its popularisation. The third regime (1980–2015) and its further reduction of annual variability in the report of new chemicals results from a larger community whose output is further regularised, not only by academic institutions and the industry, as in the second regime, but also by the frequent assistance of computerised processes, which reduce the times of synthesis, separation and characterisation of new substances. In this case, time series analysis methods allowed for detecting a remarkable historical trend, that is the stable growth rate of the chemical space, and also contingencies, gauged through the annual variability of the signal. Time series analysis also allows for predictions. In the case of the expansion of the chemical space this amounts to estimate the number of substances in a given future year. On the basis of the variability of the time series signal, we have observed that even if the third regime enjoys the lowest variability between 1800–2015, this variability is still too high to venture on accurate estimations, at least if we use only data from the most recent year to predict the next one. Nevertheless, if future advances in chemistry reduce the annual output of new substances even more, time series analysis methods will turn instrumental in determining the future size of the chemical space.

When applying time series analysis methods to data of chemical knowledge, the temporal differences of the data, discussed in Chap. 4, play an important role. For example, publications before the digital era had regular discrete outputs ranging from annual to weekly, a periodisation that is becoming more continuous with the advent of preprint servers and with the online publication of papers right after acceptance, while proofs are edited.[9] These changes are to be taken into account not only for the semiotic and the social system but also for the material system, as it implies that records of use and production of substances and reactions are also becoming continuous.[10]

[9] An instance of this approach to a continuous periodisation was recently taken by the journal *Proceedings of the National Academy of Sciences of the United States of America*, for instance [340].

[10] Inside the material system there is a further turn to continuity, since because of the ongoing robotisation of chemistry, the former reaction times ruled by our circadian clocks [200] are now determined by no-starving-no-sleeping machines.

6.4 Statistical Physics Methods

Statistical physics has developed an impressive arsenal of methods for analysing structures and data with stochastic features. Here, we can discuss only the possible applications of some sample of those methods.

From a hypergraph perspective, chemical knowledge and its evolution can be modelled as a multilayer (hyper)network, where (hyper)networks of the three systems of chemical knowledge interact with each other. Statistical physics has developed formalisms to model these networks at the level of graphs [314, 341, 342] and recent approaches move in the more general setting of hypergraphs [343], as already indicated in Sect. 6.1. Other relevant approaches of statistical physics include non-equilibrium phase transitions studies and synchronization and other coordination patterns of network dynamics [313, 344–347]. In this setting, physicists, epidemiologists, computer scientists and social scientists have modelled contagion processes, to study the diffusion of viruses, knowledge and innovations [348]. Current settings involve spreading dynamics in networks with objects driven by external factors such as human behaviours [349].

A potential application of these methods involves modelling chemical technology appropriation. For example, the driving forces behind the spread of a new catalyst or a chemical reaction. The needed information to set up these models is at hand, for instance the size of the community at a given historical time, the reports of use of those technologies and the size of the community no longer using the given technology. The mathematical settings are also well-established, as in infection disease transmission models [350].

The study of crises on the chemical knowledge and its systems is amenable to modelling using percolation settings, for example to determine network robustness to attacks [312]. These methods may shed light on the structural conditions making knowledge stable or rather the conditions facilitating periods of low innovation. Percolation may be used to pinpoint important chemical concepts facilitating the connectivity, flow and growth of chemical knowledge, which constitute mandatory concepts of any chemistry curriculum aiming at incorporating the most relevant aspects of the chemistry of its time.[11] These methods can also assist the chemical industry and the synthetic community in highlighting the effects of a shortage of fundamental substances such as acetic anhydride, the most used substrate since the 1940s [4]. They are instrumental also for taking actions caused by the sudden demand of particular chemicals such as hydroxychloroquine [352] or remdesivir [353] in the early days of the COVID-19 pandemic. Likewise, in combination with curvature measures over reactions, crucial chemical reactions for reaching new corners of the chemical space can be detected.

A straightforward application of these methods is, for instance, analysing how random deletion of chemical substances or reactions affects the structure of the chemical space.

[11] Interesting thoughts about how chemistry curricula should cope with the rapid growth of chemical knowledge are found in [351].

Likewise, one can study the stability of the social system if key individuals or hot spots of academic training are removed. A similar analysis for the semiotic system may indicate the relative importance of concepts, theories and even communication channels for the stability and evolution of chemical knowledge.

The successful application of statistical physics methods depends on the selection of the relations to study and on the dynamical details to consider. For example, the importance of the social system depends on the regarded social relations, in such a way that individual interactions may be not that relevant for chemical knowledge at large, whereas institutional relationships could be crucial. Likewise, the details of the methods to use should depend on the time scales and durations, for instance by building probabilities of social relations upon life expectancy or upon historical trends of enterprise mergers and splittings.

6.5 Agent-Based Modelling

Complex systems feature emergent phenomena arising from interactions of individual objects. Agent-based models incorporate techniques from statistical physics and concepts from network theory to model simple local interaction patterns and to understand the emerging global complex dynamics. They thus constitute a suitable framework to model emergent phenomena, where besides simple rules of interaction among objects (agents), complex object behaviour such as learning, rationality and adaptation can be incorporated.[12] Agent-based models allow for different coarse-graining levels of description. For instance, in the above example of the effect of the social system upon evolution of chemical knowledge all description levels may coexist in the model, such as persons, societies, educational systems or further objects of the social system.

As the name indicates, agent-based models look at interactions between agents, and they want to recreate or predict global patterns emerging from local activities. The fact that human agents enjoy some degree of freedom whose instantiations and effects cannot always be predicted in detail is translated into a certain randomness of the behaviour of the simulated agents. Sometimes, the individual variations then simply average out at the collective level. In other cases, correlations between agents may induce global effects, or random fluctuations may cause a transition into a different dynamical regime. While much of this is not yet understood in all detail, and the reproducibility of simulation results is also an issue, agent-based models can be a great tool for understanding the emergence of collective patterns, or more abstractly, transitions between micro- and macro phenomena. That is, for instance, the creation of scientific journals and its effect upon chemical knowledge, or the introduction of new reaction conditions and the new corners of the chemical space they open with their associated knowledge. The micro-macro

[12] The behaviour of (artificial) agents may include memory or hysteresis, path-dependence, non-Markovian behaviour, or temporal correlations [354].

understanding brought by agent-based models is a central issue not only for economics (see e.g. [355] for agent-based approaches), but also for many other domains in the social sciences. In particular, agent-based models have been used to understand reasons for collapses of societies and cultures (see for instance [356, 357]).

Agent-based models are instrumental for undertaking the estimations discussed in this document, which are based on previous knowledge of the evolution of chemistry. For example, the flow of ideas and concepts or even of molecular fragments and scaffolds. These models could be used to foresee the effects of an overpopulation of highly qualified chemistry man-power, noting that wide-spread overqualification of a substantial part of a population has been associated to periods of social instability [358]. Likewise, these models might be useful for extending classical economic models and estimating the future prices of chemicals based on the dynamics of their introduction to the community and on their early uses in scientific publications.[13] In general, these estimations may cast light on the future of chemical knowledge if the current features of its constitutive systems are kept or if these features follow the evolution of those of the last regime, assuming regimes exist.

Besides allowing for estimations, agent-based models become relevant to test different competing narratives on the history of chemistry. For example, whether the Lavoisian "revolution" was actually a disruptive and foundational event, or whether chemistry was already a discipline before Lavoisier [275, 360–362]. These models equip historians with a tool to convey rigorous, evidence-based narratives of what occurred in the past and how those events shaped today's chemistry.[14] They can also be used to explore the possible outcomes of alternative events in the past.

While theoretically the number of variables in such models can be arbitrarily large, in practice this may lead to problems of computational scales or overfitting. These latter problems occur when algorithms produce outputs that depend too strongly on the input data, and then do not generalise well and fail to estimate future observations [365]. In general, a compromise between the number of variables, agents and computational runs must be found to attain satisfactory agent-based models with straightforward interpretations [366].

Any scheme that tries to predict human behaviour may be confronted with the problem that real humans themselves try to predict the future and adjust their behaviour on the basis of those predictions. Predictions can then become self-fulfilling or self-defeating, as we regularly see, for instance, when demoscopes try to predict the outcome of an election. While the prediction may turn out to be correct in the first case, it does not capture the reason why it turns out be correct, because it will be fulfilled not because it is based on the

[13] An early work in this direction is [359], where prices of chemicals were modelled as a function of their use. Here agent-based models may refine these models by regarding the behaviour of the chemical market and the interactions of the agents, which may change their behaviour in response to the dynamics of the market.

[14] This comment is based on a plea to use agent-based models in archaeology [363]. A recent paper also discusses how in fields like economics, the past is being used to shed light on contemporary life by using statistical methods [364].

right assumptions and the right methodology, but simply because people believe that it is correct. This issue already emerged in Chap. 1 when we hinted at the possibility that the growth rate of chemical output is so stable because funding agencies and other institutions believe that this is the growth rate that should be and can be achieved. For agent-based models, or competing schemes for that matter, it is difficult to capture this reflexive aspect. When agent-based models, however, are not used for predictions, but rather for post-hoc explanations of observed phenomena, this problem may no longer be so severe, because past behaviour can no longer be adapted, even though the funding agency example would still be pertinent.

6.6 Language Theory

In previous chapters we discussed the importance of setting the semiotic system as an evolving language, as well as how chemical reactions can be understood as grammars acting upon substances. Therefore a powerful theory for the evolution of chemical knowledge is formal language theory, which provides a mathematical description of language and grammar [367]. In general, a language is formed on an alphabet, whose characters are combined into strings to produce sentences. The rules to come up with formally correct sentences of a language are encoded in its grammar.

As the evolution of chemistry requires adjusting the historical study according to the evolving ontology of chemistry, the study of the material system not only involves the level of constitution of substances, proper for the transition between the eighteenth and nineteenth centuries, but the level of molecular structures proper for the second half of the nineteenth century and for current times. A suitable model for chemical reactions at this level is that of graph rewrite rules (grammars) operating over molecular structures [368, 369], where atoms of substrates are mapped to atoms of products. This mapping is driven by a rule indicating which bonds to break and to form. Such a rule is proper for each reaction type, for example every Diels-Alder reaction requires a diene and a dienophile as part of the substrates and the formation of an unsaturated six-member ring. The rule of a Diels-Alder reaction then involves a grammar to rewrite the graph of starting materials into the graphs of products. Endowed with this formalism, grammars of chemical combination are translated into algorithms able to act upon the chemical space to foresee the future of chemistry based upon known grammars. In a temporal scale, the evolution of these grammars can be explored and can be combined with data from the social and the semiotic systems to assess the effect of those systems upon the evolution of the material system and upon the selection of chemical grammars over the time.

The preceding example of applying grammars to molecular structures works within a particular semiotic framework. At another level, the effect of the social and material system upon the selection of the grammars of the semiotic system of chemistry can be quantified. In this setting the objects of the semiotic system are the alphabet and their combinations are ruled by possible grammars that are influenced by social and chemical factors. Still at the semiotic level, language theory may shed light on the evolution of chemical discourses. What is the pool of grammars in a given period allowing chemical discourses to be accepted by the community? Here, the transition from the phlogiston chemistry of Georg Ernst Stahl (1659–1734) to that of Lavoisier comes to mind. How often do these grammars change? To which extent the current grammar of chemistry is different from that of the eighteenth century? Solving these questions is not only of utter importance for historical analyses but for the future of chemistry. If the interplay of chemical grammars with those of other disciplines is taken into account, language theory turns out to be also central for interdisciplinary studies of science.

The current theory of formal languages is based on human and programming languages and seeks to abstract their properties. It is conceivable that the grammar of chemistry, in particular when we consider the wider setting just outlined, may reveal features not present in either of those types.

6.7 Formal Concept Analysis

An important topic in the evolution of chemistry is the evolution of chemical concepts and the flow of ideas over time. This entails developing computational schemes to identify concepts and track their dynamics. A suitable method to detect those concepts, of proven use in chemistry [370, 371], is Formal Concept Analysis (FCA) [372], a mathematical technique based on a context involving objects, attributes and their relations. It detects concepts present in the given context, which are characterised by gathering objects holding a particular subset of attributes in a closed fashion. For example, the concept of chemical element in Lavoisian times was regarded as a set of substances with no decomposition reactions, which has evolved to a set of material species holding chemical equivalence, atomic number and lifetimes of at least 10^{-14} s [371].[15] As contexts can be defined for particular periods of time, the evolution of concepts boils down to analyse the concepts associated to the contexts of those periods, as exemplified with the evolution of the concept of chemical element [371].

[15] From a wider perspective, chemical elements can be thought of as pre-Lavoisian abstract bearers of chemical properties such as combustibility or acidity. As such, elements were no material objects, therefore not amenable of experimental treatment. They were rather concepts used to explain properties of material substances. After Lavoisier, elements became the end-points of chemical analysis, therefore affordable by experimental means.

The evolution of the concept of chemical element shows how an abstract concept turns material and how it evolves to a dematerialised one, although not yet abstract. This, therefore leads to discussions about the dematerialisation of contemporary chemistry. In the particular case of the superheavy elements, their half lives leave little room for chemical experiments [39].

By applying FCA to the social system of chemistry, interesting concepts may arise, for example gathering together social actors such as chemists and institutions, which are characterised by economic ties or by regions of the chemical space collectively explored. Concepts for the whole system of chemical knowledge involve the objects of the three systems compounding chemical knowledge and the possible relationships they may establish. In this sense, FCA becomes also a further framework to characterise, and further study, chemical knowledge and its evolution.

FCA, beyond finding concepts, allows for detecting relationships among them, which are used to detect meaningful inferential rules of the given context and its conceptual structure. Inferential rules are statements of the sort "every object with attributes $a, b, c, \ldots$ also has the attributes $x, y, z, \ldots$" [372]. Some examples of inferential rules related to the concept of acid, for instance, are: every chemical producing a sour taste contains oxygen, or every substance that dissociates in water by protonating it, also transfers the proton to a base. FCA inference rules may be also flexible by indicating, for example, "for 70% of the substances of the context it is true that protonating water implies transferring protons to bases." Finding these sorts of inference rules for different contexts of chemical knowledge, and in different periods, are of central importance for understanding the evolution of chemical knowledge. In general, FCA arranges concepts by the degree of subsethood of objects and attributes, producing different order structures, actually lattices. The application of FCA to different periods of the evolution of the chemical knowledge may lead to different conceptual structures, which can be compared in the search of isomorphisms or conceptual paths over the history. Here one may consider whether the conceptual structure of chemistry in a particular period was similar to that of other sciences and how such a similarity is evolving.

It has been shown that concepts exhibit some geometric properties such as monotonicity, continuity and convexity, which are related to causality, action control and generalisation and classification of knowledge [373]. This, applied to the concepts detected through FCA, constitutes a promising line of research.

Despite the applicability of FCA to studies of chemical knowledge, the analysis of evolving concepts requires a careful selection of attributes and objects. If many attributes are taken into account, the interpretation of concepts turns cumbersome. Other problems may arise from the discretisation of continuous attributes, which is a requirement for the technique.

6.8 Text Analysis and Computational Semiotics/Semantics

Scientific publications in chemistry constitute a corpus upon which the evolution of concepts and thematics can be analysed using text analysis tools. The aim is to use these techniques to convert language data to machine understandable form and back again [374]. In contrast to formal language theory, text analysis also aims at capturing semantic aspects.[16] Text analysis can therefore become a tool that complements the linguistic approach to chemical reactions as rewrite rules, for example in the evolution of the material system, where information on reaction conditions and the methodological settings is incorporated in reaction descriptions. This is a relevant addition, not only for the sake of completeness, but because it has been found that scientific publications combining text and visualisations may convey different stories in the two communication channels [377]. A further application of these tools to the material system involves regarding reactions as sentences and the objects of study are then their words (substances). In this setting, the co-occurrence and further replaceability of substances in sentences may shed light on chemical similarity, which may lead to a classification of substances based upon their role in chemical reactions, rather than on their molecular structure. This setting is computationally amenable, as substances need not be encoded at the molecular level but simply through identificators, such as database ids or Chemical Abstracts Service numbers. This allows for studying even historical periods where uncertainty in molecular structure was the rule, or when the concept was not yet developed.

An important text analysis tool is text summarisation [378], which is instrumental to analyse, for instance, minutes of industrial committees. Likewise, to systematically build up summaries of chemical literature lacking them such as old documents or scientific papers with no "abstract" component. This technique, therefore becomes a normalisation tool for comparing chemical literature from different time periods. Text analysis tools are also important to quantify the size of the chemical community with particular interests or associated to particular concepts, themes or topics. This can be done by exploring how often particular terms appear in the literature and how those terms are associated with others. For example, how frequent distillation is associated with diseases, or more recently, how often "global warming" appears in organic, inorganic or analytical chemistry documents. In this setting the unveiled thematics are associated to authors, allowing for the analysis of the dynamics of concepts.

As chemical knowledge is encoded in a large corpus, computational methods are required, which complement traditional bibliometric approaches. Computational semiotics methods may fill this gap. Some of these tools are devised to analyse persuasion, for instance, through opposition analysis [379, 380], which might cast light on the workings of innovation in chemistry and the effectiveness of communication channels to attain acceptance of ideas, theories, methods, substances and reactions, for instance. These techniques allow for determining, for instance, how influential ideas in chemistry, such

[16] The distinction between semiotics and semantics is somewhat controversial in the literature. But many semioticists may agree (e.g. [375, 376]) that semantics is concerned with the meaning and the reference of (linguistic) expressions, while semiotics is the general theory of signs. We are here concerned with the reference and the meaning of chemical writings, but also in the semiotic relations inside and between texts. Moreover, as will be explained below, there might exist analogies between the general structure of how meaning is assigned to words and sentences and how the meanings of different words are related with the structure of chemical space.

as the Lavoisian system for chemistry, were initially perceived by the community and later on assimilated;[17] or how competing theories gain or lose adepts and how these dynamics are related to others opinions and to further scientific evidence.[18] Another interesting tool is provided by semantic vector spaces models, which allow for discovering combinatorial patterns of signs leading to inferring complex meaning structures such as paradigmatic relations, which takes into account the degree of replaceability of two signs in a corpus in such a manner that two signs are semantically similar if they have similar combinatorial patterns [381]. The fundamental linguistic distinction of Ferdinand de Saussure (1857–1913) [382] between syntagmatic and paradigmatic relations seems also very relevant for chemistry. Words that can be combined into phrases stand in a syntagmatic relation, whereas those that can replace each other in a phrase have a paradigmatic relation. In *the big building*, *big* and *building* are syntagmatically related, whereas *building* and *house* are paradigmatically related, as we can replace the phrase by *the big house*. Similarly, chemical substances can be combined in reactions, or can replace each other as substrates. This is quite relevant, for example, for understanding the logic behind the periodic system and how this system emerged historically [216]. Therefore, we see a strong potential for computational semantic methods in chemistry. Moreover, such methods are instrumental for detecting similarity of chemical social structures, for instance. Likewise, they can cast light on similar concepts over the history of chemistry.

Some of the limitations of these methods involve determining new features of combinations to produce summaries, for instance. Other problems arise with ambiguous sentences or words (with more than one meaning) or with synonyms, which may distort the final results [383]. These problems do not seem insurmountable, however.

6.9 Machine Learning

Machine learning (ML) is a large and rapidly growing field that incorporates many of the methods described in the previous chapters. It includes supervised, unsupervised and reinforcement learning algorithms. The first ones are mainly used for prediction, where the algorithm provides a function used for estimations, which it learns from a large dataset. Once such an algorithm is trained on some data set, its predictive abilities can then be tested on a new data set. Examples of those algorithms are regression techniques. Unsupervised learning entails detecting patterns from a dataset. Cluster analysis is one

[17] This question has been already addressed, for instance in [212] using traditional methods of the history of science. Nevertheless, a computational approach to the subject may be used to contrast the results of both methodologies and, perhaps, to shed light on aspects not yet considered.

[18] Traditional bibliometric methods, for example by counting frequencies of citations, do not easily capture this information, as citations do not inform about the intention of the citations, that is, whether it is to support and idea or to criticise it.

of these algorithms. Reinforcement learning algorithms make decisions based on rewards and maximise the reward; many agent-based models belong to this category.

ML approaches have found applications in most areas of knowledge, for example in sociology they have been used to estimate outcomes of deadly conflicts [384, 385], where the results are used to better understand the underlying processes and to sharpen the available theories [386]. The social system of chemistry and the chemical knowledge could benefit from these approaches, for example to understand crises in chemical knowledge and their relationship with social driving forces. In terms of classifications, sociologists have used ML, for instance, to infer the social impacts of different public policies [387]. In chemistry, similar approaches could be used to estimate the impact upon chemical knowledge of political and environmental decisions, for instance. Similar algorithms have found use in semantics, for example for analysing media discourses, which also combine text analysis techniques [388].

Chemistry has also benefited by incorporating ML algorithms, which have allowed scientists to speed up the unveiling of the chemical space [389, 390], predict properties of substances, optimise synthesis plans [391] and review the chemical literature [123], among other applications [69, 392, 393]. ML methods are a suitable computational tool to, for example, determine the grammars of combination over the history of chemistry and to determine how their population has evolved over the time as influenced by material, semiotic and social driving forces. They could also become instrumental in the classification of chemical reactions according to their rewrite rules and further reaction data and metadata available in large repositories. This may be used to foresee the rise of new reaction classes and to detect the suitable conditions facilitating those innovations.

ML algorithms strongly depend on the quality and diversity of the training datasets [394]. Although chemistry is not short of data for carrying out the studies presented here, multidisciplinary efforts are required to generate large curated and diverse datasets to avoid biased results. A further limitation of ML methods is the risk of overfitting (discussed in Sect. 6.5) and the lack of approaches to gauge causal relationships in the datasets [395]. Nevertheless, several approaches are currently explored to meet this challenge. ML techniques can be quite good in detecting correlations in large data sets. The task of finding structure in data sets without a guiding theory may lead to a systematic reorganisation of scientific practice and pose new challenges for the formal sciences [396]. Often such correlations had not been expected, but they can turn out to be very useful for predictions. However, correlation does not imply causation, and it is usually the scientific aim to infer the latter. Building upon the theory of causality in Bayesian networks developed by Judea Pearl (1936–) [397], a powerful method for detecting causal relations consists in the analysis of the propagation of perturbations in coupled time series [398]. When some perturbation inserted at A subsequently produces an effect at B, then one may suspect that there is a causal influence from A to B. This influence may, however, be indirect, for instance mediated by C, and therefore a more refined analysis of the data may be required.

The method of choice in current ML are deep neural networks (DNNs). For reasons that are not yet really well understood [399], their hierarchical organisation makes it possible to extract high level relationships between data. Often, they have vastly more parameters (the connection strengths between all formal units, the so-called neurons) than there are data points available. Standard wisdom would suggest that this may cause a terrible overfitting of the data, but apparently, there is some intrinsic regularising mechanism at work that avoids such overfitting. At current, it is only poorly understood why this is so and how it works in detail [400], but DNNs are remarkably successful at all kinds of tasks. In particular, they are superior to other ML schemes when the structure in the data is unclear or opaque.

The methods discussed here are not to be taken as "canned" tools of straightforward application. Without exception, they require previous knowledge of the data and of the processes leading to them. In particular, they have typically been developed for other purposes and types of data and therefore need to be adjusted for use in chemistry and/or the history of science. This, therefore, requires joint work among chemists, historians, linguists and other scientists and scholars. The right application of the methods discussed here requires permanent discussions and joint work with mathematicians, who not only know the details of these methods but may adapt them or develop further ones to solve the questions posed by the complexity of chemical knowledge.

7 Conclusions and Outlook

Chemistry is not only of interest for chemists, but rather for the society at large, as the output of this discipline generates both welfare and hazard for our civilisation. Therefore, understanding chemical knowledge and its evolution is of central importance. In this book we introduced a formal setting for chemical knowledge as a complex dynamical system resulting from the mutual interaction of the social, semiotic and material systems of chemistry. We provided mathematical formalisms and methods to describe the dynamics of these systems, from which chemical knowledge emerges. Likewise, we presented the different data sources to carry out this research programme and discussed their reaches and challenges. We have shown how different disconnected parts of this programme have already started. It is the aim of this book to bring unity to those efforts by providing formal structures transversal to them.[1]

Studying the evolution of chemical knowledge requires developing understanding among chemists, historians, sociologists, linguists, physicists, computer scientists and mathematicians, among others. Besides gathering and curating data, and selecting the right analytical methods, this interdisciplinary engagement is the current major challenge for these studies. This can be overcome by breaking down the current walls separating disciplines through the creation of educational and institutional research settings facilitating the interaction of interdisciplinary teams. Although a couple of these spaces already exist, such as the Santa Fe Institute, the Complexity Science Hub Vienna and the Max Planck Institute for Mathematics in the Sciences, more institutions of these sorts are needed. Overcoming the interdisciplinary challenge also requires an adjustment of the funding system of research, beyond the traditional grant-funded initiatives at individual universities or research institutes. A visionary funding system should seek and stimulate

[1] A recent call for a systems view of chemistry was reported by Anastas [401].

J. Jost, G. Restrepo, *The Evolution of Chemical Knowledge*, Wissenschaft und Philosophie – Science and Philosophy – Sciences et Philosophie,
https://doi.org/10.1007/978-3-031-10094-9_7

interdisciplinary projects that address the issues that we have described and worked out. Ecology, for instance, confronted a similar challenge in the 1990s and came up with an innovative solution through the launching of the National Center for Ecological Analysis and Synthesis, mainly funded by the National Science Foundation [363]. Chemistry being the highly productive scientific discipline it is, which shapes and creates the world's material resources, should it not follow the steps of ecology?

At the educational front, the study of the evolution of chemical knowledge involves building up interdisciplinary ways of thinking. Successful results in this direction have been obtained by the Santa Fe Institute through its on-site and virtual courses and schools [402]. Although the advantages of interdisciplinary thinking have been highlighted [403], chemistry has only devised pedagogical interdisciplinary strategies to specific local chemistry subdisciplines.[2] Therefore, efforts encompassing the entire chemistry community are mandatory. In the meantime, interdisciplinary thinkers are needed, which in the realm of chemistry do not abound, perhaps as a consequence of the historical disciplinary aim of producing new substances and chemical reactions [404].[3]

Chemistry has a tradition of secrecy that is tied to its historical commercial and industrial links. Evolution of chemical knowledge studies, besides breaking down academic barriers, need to overcome social ones, or at least negotiate them in such a manner that industrial and proprietary information can be used for the sake of chemical knowledge. This is not an utopia in our contemporary society and especially in the social system of chemical knowledge. The disciplinary advantages of openness and data sharing are well documented [405, 406], and chemistry is actually moving in the direction of open data, open source, and open standards [407]. While freely available data are at hand, academic agreements can be signed between the industry and research institutions, where the commercial interests of one part are guaranteed while providing access for the purposes of this research programme. There is in fact evidence of such agreements, for example the one between Elsevier, owner of Reaxys©, and our research group and the different possibilities Web of Science [300] and Dimensions [301] offer to dump data and to conduct analyses.

As Murray Gell-Mann pointed out, besides working on the specialities of each discipline it is worth giving "a crude look at the whole" [408]. This has been also the subject of recent discussion in archaeology, where the importance has been recognised of devoting much more efforts to accessing, analysing, and comparing different data sets to come up with explanations and insights that could never emerge from the analysis of individual projects [363]. We believe this is the situation of chemistry, with a rich past of patterns to be unveiled, which are driven by several internal and external forces. The

[2] Examples of these are found in several papers published over the years in the Journal of Chemical Education.

[3] This has been also argued as the cause of the lack of interest on the philosophical issues of chemistry [6].

available formalisms, data and computational facilities make this moment in the history of chemistry ripe to undertake full understanding of the evolution of chemical knowledge, with its implications for the narratives of the past and, above all, for the future of chemistry and our civilisation.

References

1. Joyce GF (1997) Evolutionary chemistry: getting there from here. Science 276:1658–1659
2. Reinhardt C (2018) Introduction. Isis 109:559–564
3. Reaxys (2022) an expert-curated chemistry database. https://www.elsevier.com/solutions/reaxys. Accessed 13 Jan 2022
4. Llanos EJ, Leal W, Luu DH, Jost J, Stadler PF, Restrepo G (2019) Exploration of the chemical space and its three historical regimes. Proc Natl Acad Sci 116:12660–12665
5. de Solla Price DJ (1963) Little science, big science. Columbia University Press, Columbia
6. Schummer J (2006) The philosophy of chemistry: from infancy toward maturity. Boston Studies in the Philosophy of Science. Springer, Berlin, pp. 19–39
7. Chang SH (2018) A pilot study on the connection between scientific fields and patent classification systems. Scientometrics 114:951–970
8. Friedman M (2011) Kant's construction of nature. Cambridge University Press, Cambridge
9. Van Brakel J (2006) Kant's legacy for the philosophy of chemistry. Boston Studies in the Philosophy of Science. Springer, Berlin, pp 69–91
10. Kant I (1995) Opus postumum. Cambridge University Press, Cambridge
11. Restrepo G (2013) To mathematize, or not to mathematize chemistry. Found Chem 15:185–197
12. Carrier M (1990) Kants Theorie der Materie und ihre Wirkung auf die zeitgenössische Chemie . Kant-Studien 81:170–210
13. Shannon CE (1948) A mathematical theory of communication. Bell Syst Tech J 27:379–423
14. Putnam H (1975) The meaning of meaning. Philosophical papers 2. University of Minnesota Press, Minneapolis
15. Jost J (2020) Biological information. Theory Biosci 139:361–370
16. Gettier EL (1963) Is justified true belief knowledge? Analysis 23:121–123
17. Mannheim K (1929) Ideologie und Utopie. Cohen; English translation: ideology and utopia; an introduction to the sociology of knowledge, 1936. F. Cohen, Bonn
18. Berger P, Luckmann T (1966) The social construction of reality. A Treatise in the Sociology of Knowledge. Anchor Books, Garden City, NY
19. Renn J (2020) The evolution of knowledge: rethinking science in the anthropocene. Princeton University Press, Princeton
20. Rothbart D (2004) Modeling: gateway to the unknown: a work by Rom Harre. ISSN. Elsevier Science, Amsterdam
21. Najemnik J, Geisler WS (2008) Eye movement statistics in humans are consistent with an optimal search strategy. J Vis 8:4–4

J. Jost, G. Restrepo, *The Evolution of Chemical Knowledge*, Wissenschaft und Philosophie – Science and Philosophy – Sciences et Philosophie,
https://doi.org/10.1007/978-3-031-10094-9

22. Lynn CW, Bassett DS (2020) How humans learn and represent networks. Proc Natl Acad Sci 117:29407–29415
23. Johnson-Laird PN (2006) How we reason. Oxford University Press, Oxford
24. Spohn W (2012) The laws of belief: ranking theory and its philosophical applications. Oxford University Press, Oxford
25. Krakauer DC (2011) The star gazer and the flesh eater: elements of a theory of metahistory. Cliodynamics 2:82–105
26. Schummer J (1998) The chemical core of chemistry I: a conceptual approach. Hyle 4:129–162
27. Locke J (1690) An essay concerning humane understanding. Thomas Basset, London
28. Noyes WA (1911) The elements of qualitative analysis, sixth edition revised in collaboration with G. McP. Smith. Henry Holt, New York
29. Brock WH (1993) The Norton history of chemistry. W. W. Norton & Company, New York
30. Timmermans J, Oesper R (1963) The concept of species in chemistry. Chemical Publishing Company, Revere
31. Hoffmann R, Laszlo P (1991) Representation in chemistry. Angew Chem Int Ed Engl 30:1–16
32. Jost J, Scherrer K (2014) Information theory, gene expression, and combinatorial regulation: a quantitative analysis. Theory Biosci 133:1–21
33. Pomerantseva E, Bonaccorso F, Feng X, Cui Y, Gogotsi Y (2019) Energy storage: the future enabled by nanomaterials. Science 366:eaan8285
34. Huang W, Ding Z, Wang C, Wei J, Zhao Y, Purnawali H (2010) Shape memory materials. Mat Today 13:54–61
35. Leicester H (1971) The historical background of chemistry. Dover books on chemistry, physical chemistry, biochemistry. Dover Publications, Mineola
36. International society for the philosophy of chemistry (2021) https://sites.google.com/site/socphilchem/. Accessed 7 Apr 2021
37. Society for Philosophy of Science in Practice (2021) https://www.philosophy-science-practice.org/. Accessed 7 Apr 2021
38. Falkenburg B (2007) Particle metaphysics: a critical account of subatomic reality. Springer Science & Business Media, Berlin
39. Restrepo G (2019) Challenges for the periodic systems of elements: chemical, historical and mathematical perspectives. Chem Eur J 25:15430–15440
40. Brock WH (2016) The history of chemistry: a very short introduction. Very short introductions. Oxford University Press, Oxford
41. Lakatos I (1976) Proofs and refutations: the logic of mathematical discovery. Cambridge university press, Cambridge
42. Nye MJ (1994) From chemical philosophy to theoretical chemistry: dynamics of matter and dynamics of disciplines. University of California Press, Berkeley
43. Lenoir T (1997) Instituting Science: the Cultural Production of Scientific Disciplines. Writing science. Stanford University Press, Redwood City
44. Bensaude-Vincent B, Deborah Van Dam B, Stengers I, Van Dam D, van Dam D (1996) A history of chemistry. Religion and politics. Harvard University Press, Cambridge
45. Hessen B (2009) The social and economic roots of Newton's principia. Springer Netherlands, Dordrecht, pp 41–101
46. Grossmann H (1935) Die gesellschaftlichen Grundlagen der mechanistischen Philosophie und die Manufaktur. Z Soziol 4:161–231
47. Grossmann H (1987) The social foundations of mechanistic philosophy and manufacture. Sci Context 1:129–180
48. Fleck L (1979) The genesis and development of a scientific fact. University Chicago Press, Chicago. Translated by Trenn T, Merton R

49. Restrepo G (2020) Das periodische System und die Evolution des chemischen Raums. Nachr Chem 68:12–15
50. Grethe G, Blanke G, Kraut H, Goodman JM (2018) International chemical identifier for reactions (RInChI). J Cheminf 10:22
51. Johnson MA, Maggiora GM (eds) (1990) Concepts and applications of molecular similarity. Wiley, Hoboken
52. Roy K (ed.) (2017) Advances in QSAR modeling. Springer, Berlin
53. Huisgen R (1986) Adolf von Baeyer's scientific achievements — a legacy. Angew Chem Int Ed Engl 25:297–311
54. Maier H (2015) Chemiker im "Dritten Reich": die Deutsche Chemische Gesellschaft und der Verein Deutscher Chemiker im NS-Herrschaftsapparat. Wiley, Hoboken
55. Johnson JA (2017) Between nationalism and internationalism: the German chemical society in comparative perspective, 1867–1945. Angew Chem Int Ed 56:11044–11058
56. Johnson JA (1989) Hierarchy and creativity in chemistry, 1871–1914. Osiris 5:214–240
57. Lipinski CA, Lombardo F, Dominy BW, Feeney PJ (1997) Experimental and computational approaches to estimate solubility and permeability in drug discovery and development settings. Adv Drug Deliv Rev 23:3–25
58. Anastas P, Warner J (2000) Green chemistry: theory and practice. Oxford University Press, Oxford
59. Alam MT, Olin-Sandoval V, Stincone A, Keller MA, Zelezniak A, Luisi BF, Ralser M (2017) The self-inhibitory nature of metabolic networks and its alleviation through compartmentalization. Nat Commun 8:16018
60. Graßhoff G, May M (2003) Hans Krebs' and Kurt Henseleit's laboratory notebooks and their discovery of the urea cycle-reconstructed with computer models. Springer Netherlands, Dordrecht, pp 269–294
61. Nickelsen K, Graßhoff G (2009) Concepts from the Bench: Hans Krebs, Kurt Henseleit and the urea cycle. Springer Netherlands, Dordrecht, pp 91–117
62. Frängsmyr T, Heilborn H, Heilbron J, Rider R (1990) The quantifying spirit in the 18th century. Comparative studies on muslim societies. University of California Press, Berkeley
63. Myers KR, Tham WY, Yin Y, Cohodes N, Thursby JG, Thursby MC, Schiffer P, Walsh JT, Lakhani KR, Wang D (2020) Unequal effects of the COVID-19 pandemic on scientists. Nat Hum Behav 4:880–883
64. Day AE, Corbett P, Boyle J (2020) Is there a gender gap in chemical sciences scholarly communication? Chem Sci 11:2277–2301
65. ACS (2020) A letter from ACS president on ACS diversity, inclusion & respect commitments and activities. https://www.acs.org/content/acs/en/membership-and-networks/acs/welcoming/diversity/diversity-inclusion-and-respect-letter-august-2020.html. Accessed 23 Nov 2020
66. Kueckmann T, Compton N, Maaß F, Su X, Weickgenannt N (2020) Redefining Angewandte Chemie as an inclusive home for excellent science. Angew Chem Int Ed 59(52):23348–23349
67. Brown DG, Boström J (2016) Analysis of past and present synthetic methodologies on medicinal chemistry: where have all the new reactions gone? J. Med Chem 59:4443–4458
68. Boström J, Brown DG (2016) Stuck in a rut with old chemistry. Drug Discov Today 21:701–703
69. Gromski PS, Henson AB, Granda JM, Cronin L (2019) How to explore chemical space using algorithms and automation. Nat Rev Chem 3:119–128
70. Ke Q, Ferrara E, Radicchi F, Flammini A (2015) Defining and identifying sleeping beauties in science. Proc Natl Acad Sci 112:7426–7431
71. Klein U (2003) Experiments, models, paper tools: cultures of organic chemistry in the nineteenth century. Stanford University Press, Redwood City

72. Grudniewicz A, Moher D, Cobey KD, Bryson GL, Cukier S, Allen K, Ardern C, Balcom L, Barros T, Berger M, Ciro JB, Cugusi L, Donaldson MR, Egger M, Graham ID, Hodgkinson M, Khan KM, Mabizela M, Manca A, Milzow K, Mouton J, Muchenje M, Olijhoek T, Ommaya A, Patwardhan B, Poff D, Proulx L, Rodger M, Severin A, Strinzel M, Sylos-Labini M, Tamblyn R, van Niekerk M, Wicherts JM, Lalu MM (2019) Predatory journals: no definition, no defence. Nature 576:210–212
73. Weingart P, Taubert N (2017) The future of scholarly publishing. African Minds, Cape Town
74. Kupferschmidt K (2020) 'A completely new culture of doing research.' Coronavirus outbreak changes how scientists communicate. Science. https://www.science.org/content/article/completely-new-culture-doing-research-coronavirus-outbreak-changes-how-scientists. Publication date: 20 Feb 2020.
75. Viglione G (2020) A year without conferences? How the coronavirus pandemic could change research. Nature 579:327–328
76. Warr W (2015) Many InChIs and quite some feat. J Comput Aided Mol Des 29:681–694
77. Heller S (2017) InChI trust project director's report. Technical Report, InChITRUST
78. Bucior BJ, Rosen AS, Haranczyk M, Yao Z, Ziebel ME, Farha OK, Hupp JT, Siepmann JI, Aspuru-Guzik A, Snurr RQ (2019) Identification schemes for metal-organic frameworks to enable rapid search and cheminformatics analysis. Cryst Growth Des 19:6682–6697
79. Clark AM (2011) Accurate specification of molecular structures: the case for zero-order bonds and explicit hydrogen counting. J Chem Inf Model 51:3149–3157
80. Gell-Mann M (2017) Regularities in human affairs, chap. 4. SFI Press, Santa Fe, pp 63–90
81. Friedrich B, Hoffmann D, Renn J, Schmaltz F, Wolf M (eds) (2017) One hundred years of chemical warfare: research, deployment, consequences. Springer, Berlin
82. Molina MJ, Rowland FS (1974) Stratospheric sink for chlorofluoromethanes: chlorine atom-catalysed destruction of ozone. Nature 249:810–812
83. Hogue C (2021) There's no need to control PFAS as a class, industry scientists say. Chem Eng News 99(19)
84. Nicolaou KC, Vourloumis D, Winssinger N, Baran PS (2000) The art and science of total synthesis at the dawn of the twenty-first century. Angew Chem Int Ed 39:44–122
85. Wise MN (2004) Introduction: dynamincs all the way up. Duke University Press, Durham, pp 1–20
86. Wise MN (2011) Science as (historical) narrative. Erkenntnis (1975-) 75:349–376
87. Carrier M, Mertens R, Reinhardt C (eds) (2021) Narratives and comparisons. Adversaries or allies in understanding science? Bielefeld University Press, Bielefeld
88. Scerri E (2019) The periodic table: its story and its significance. Oxford University Press, Oxford
89. Bensaude-Vincent B (1986) Mendeleev's periodic system of chemical elements. Br J Hist Sci 19:3–17
90. Rocke A (2019) Lothar Meyer's pathway to periodicity. Ambix 66:265–302
91. Rocke AJ (1984) Chemical atomism in the nineteenth century. Ohio State University Press, Columbus
92. Leal W, Llanos EJ, Bernal A, Stadler PF, Jost J, Restrepo G (2022) The expansion of chemical space in 1826 and in the 1840s prompted the convergence to the periodic system. Proc Natl Acad Sci USA 119:e2119083119
93. Dalton J (1810) A new system of chemical philosophy, part 2. S. Russell, Manchester
94. Thomson T (1813) On the Daltonian theory of definite proportions in chemical combinations. Ann Philos 2:32–52
95. Berzelius JJ (1819) Essai sur la théorie des proportions chimiques et sur l'influence chimique de l'électricité. Mequignon-Marvis, Paris

96. Berzelius JJ (1826) Ueber die Bestimmung der relativen Anzahl von einfachen Atomen in chemischen Verbindungen. Ann Phys 83:397–416
97. Berzelius JJ (1826) Ueber die Bestimmung der relativen Anzahl von einfachen Atomen in chemischen Verbindungen. Ann Phys 84:1–24
98. Berzelius JJ (1826) Ueber die Bestimmung der relativen Anzahl von einfachen Atomen in chemischen Verbindungen. Ann Phys 84:177–190
99. Gmelin L (1843) Handbuch der chemie, 8 v. in 9. Karl Winter, Heidelberg
100. Lenßen E (1857) Ueber die Gruppirung der Elemente nach ihrem chemisch - physikalischen Character. Justus Liebigs Ann Chem 103:121–131
101. Meyer L (1864) Die modernen Theorien der Chemie und ihre Bedeutung für die chemische Statik. Verlag von Maruschke & Berendt, Breslau
102. Rex F (1997) Zur Erinnerungen an Felix Hoppe-Seyler, Lothar Meyer und Walter Hückel. Bausteine zur Tübinger Universitätsgeschichte 8:103–130
103. Meyer L (1870) Die Natur der chemischen Elemente als Function ihrer Atomgewichte. Ann Chem Pharm VII Supplementband:354–364
104. Odling W (1864) On the proportional numbers of the elements. Q J Sci 1:642–648
105. Hinrichs G (1867) Programme der Atomechanik oder die Chemie eine Mechanik der Panatome. Privately printed, Iowa City, IA.
106. Mendeleev D (2002) On the relation of the properties to the atomic weights of the elements. In: Jensen WB (ed) Mendeleev on the periodic law: selected writings, 1869–1905, chap 1. Dover, New York, pp 16–17
107. Pauling L (1929) The principles determining the structure of complex ionic crystals. J Am Chem Soc 51:1010–1026
108. George J, Waroquiers D, Di Stefano D, Petretto G, Rignanese GM, Hautier G (2020) The limited predictive power of the Pauling rules. Angew Chem Int Ed 59(19), pp. 7569–7575
109. Schummer J (1997) Scientometric studies on chemistry I: the exponential growth of chemical substances, 1800–1995. Scientometrics 39:107–123
110. Lesch JE (1990) Systematics and the geometrical spirit. University of California Press, Berkeley, pp. 73–112
111. Cassirer E, Koelln F, Koelln F, Pettegrove J (1979) The Philosophy of the Enlightenment. Beacon paperbacks. Princeton University Press, Princeton
112. Restrepo G, Villaveces JL (2011) Chemistry, a lingua philosophica. Found Chem 13:233–249
113. Jost J (2019) Leibniz und die moderne Naturwissenschaft. Springer, Berlin
114. Bertalanffy Lv (1950) An outline of general system theory. Br J Philos Sci I:134–165
115. Lambert JH (1988) Texte zur Systematologie und zur Theorie der wissenschaftlichen Erkenntnis. Fragmente einer Systematologie. Felix Meiner, Hamburg, pp 125–144
116. Wintergrün D (2019) Netzwerkanalysen und semantische Datenmodellierung als heuristische Instrumente für die historische Forschung. doctoralthesis, Friedrich-Alexander-Universität Erlangen-Nürnberg (FAU)
117. Hooker C (2011) Introduction to philosophy of complex systems: A: part a: towards a framework for complex systems. In: Hooker C (ed) Philosophy of complex systems. Handbook of the philosophy of science, vol 10. North-Holland, Amsterdam, pp 3–90
118. Crosland MP (1962) Historical studies in the language of chemistry. Harvard University Press, Cambridge
119. Döbereiner JW (1829) Versuch zu einer Gruppirung der elementaren Stoffe nach ihrer Analogie. Ann Phys 91:301–307
120. Jost J (2004) External and internal complexity of complex adaptive systems. Theory Biosci 123:69–88
121. Rescher N (1978) Scientific progress. Basil Blackwell, Oxford

122. US patent office rules that artificial intelligence cannot be a legal inventor (2020) https://www.theverge.com/2020/4/29/21241251/artificial-intelligence-inventor-united-states-patent-trademark-office-intellectual-property. Accessed 10 Jul 2020
123. Writer B (2019) Lithium-ion batteries. Springer, Berlin
124. Ley SV, Fitzpatrick DE, Myers RM, Battilocchio C, Ingham RJ (2015) Machine-assisted organic synthesis. Angew Chem Int Ed 54:10122–10136
125. Steiner S, Wolf J, Glatzel S, Andreou A, Granda JM, Keenan G, Hinkley T, Aragon-Camarasa G, Kitson PJ, Angelone D, Cronin L (2019) Organic synthesis in a modular robotic system driven by a chemical programming language. Science 363
126. Burger B, Maffettone PM, Gusev VV, Aitchison CM, Bai Y, Wang X, Li X, Alston BM, Li B, Clowes R, Rankin N, Harris B, Sprick RS, Cooper AI (2020) A mobile robotic chemist. Nature 583:237–241
127. MacLeod BP, Parlane FGL, Morrissey TD, Häse F, Roch LM, Dettelbach KE, Moreira R, Yunker LPE, Rooney MB, Deeth JR, Lai V, Ng GJ, Situ H, Zhang RH, Elliott MS, Haley TH, Dvorak DJ, Aspuru-Guzik A, Hein JE, Berlinguette CP (2020) Self-driving laboratory for accelerated discovery of thin-film materials. Sci Adv 6
128. Schummer J, Bensaude-Vincent B, Van Tiggelen B (2007) The public image of chemistry. World Scientific, Singapore
129. Mehlich J, Moser F, Van Tiggelen B, Campanella L, Hopf H (2017) The ethical and social dimensions of chemistry: reflections, considerations, and clarifications. Chem Eur J 23:1210–1218
130. Schummer J (2001) Ethics of chemical synthesis. Hyle 7:103–124
131. Contakes SM, Jashinsky T (2016) Ethical responsibilities in military-related work: the case of Napalm. Hyle 22:31–53
132. Herz N, Dan O, Censor N, Bar-Haim Y (2020) Opinion: authors overestimate their contribution to scientific work, demonstrating a strong bias. Proc Natl Acad Sci 117:6282–6285
133. Peirce CS (1998) The essential Peirce, vol 2. Indiana University Press, Bloomington
134. Short TL (2007) Peirce's theory of signs. Cambridge University Press, Cambridge
135. Water (2020) https://www.etymonline.com/word/water. Accessed 25 Nov 2020
136. Agua (2020) https://etimologia.wordpress.com/2007/02/26/agua/. Accessed 25 Nov 2020
137. Hoffmann R, Hoffmann F (1995) The same and not the same. George B. Pegram lecture series. Columbia University Press, New York
138. Weininger SJ (1998) Contemplating the finger: visuality and the semiotics of chemistry. Hyle 4:3–27
139. Chang H (2011) The persistence of epistemic objects through scientific change. Erkenntnis 75:413–429
140. Sadownik JW, Otto S (2014) Systems chemistry. Berlin, Heidelberg, pp 1–3
141. Tu K, Gusak A (2014) Kinetics in nanoscale materials. Wiley, New York
142. Taape T (2014) Distilling reliable remedies: Hieronymus Brunschwig's liber de arte distillandi (1500) between alchemical learning and craft practice. Ambix 61:236–256
143. Kuhn T (1962) The structure of scientific revolutions. University Chicago Press, Chicago
144. Planck M (2001) Max Planck: Vorträge, Reden, Erinnerungen. Springer, Berlin
145. Planck M (1949) Scientific autobiography, and other papers. Williams & Norgate, London
146. Azoulay P, Fons-Rosen C, Graff Zivin JS (2019) Does science advance one funeral at a time? Am Econ Rev 109:2889–2920
147. Kildebæk Nielsen A, Strbanova S (eds) (2008) Creating networks in chemistry. Special Publications. The Royal Society of Chemistry, London
148. Andersen JP, Hammarfelt B (2011) Price revisited: on the growth of dissertations in eight research fields. Scientometrics 88:371–383

149. Sugimoto CR, Li D, Russell TG, Finlay SC, Ding Y (2011) The shifting sands of disciplinary development: analyzing north american library and information science dissertations using latent dirichlet allocation. J Am Soc Inf Sci Technol 62:185–204
150. Nielsen AK, Štrbáňová S (2008) Chapter 15 creating networks in chemistry – some lessons learned. In: Creating networks in chemistry: the founding and early history of chemical societies in Europe. The Royal Society of Chemistry, London, pp 328–348
151. Fell U, Rocke A (2008) Chapter 5 france: the chemical society of france in its formative years, 1857–1914: disciplinary identity and the struggle for unity. In: Creating networks in chemistry: the founding and early history of chemical societies in Europe. The Royal Society of Chemistry, London, pp 91–112
152. Johnson JA (2008) Chapter 6 Germany: Discipline – industry – profession. German chemical organizations, 1867–1914. In: Creating networks in chemistry: the founding and early history of chemical societies in Europe. The Royal Society of Chemistry, London, pp 113–138
153. Knight D, Kragh H (2008) The making of the chemist: the social history of chemistry in Europe, 1789–1914. Cambridge University Press, Cambridge
154. Homburg E (2018) Chemistry and industry: a tale of two moving targets. Isis 109:565–576
155. Schummer J (2002) The impact of instrumentation on chemical species identity: from chemical substances to molecular species. Royal Society of Chemistry, London, pp 188–211
156. Morris PJT, Travis AS, Reinhardt C (2001) Research fields and boundaries in twentieth-century organic chemistry. In: Reinhardt C (ed) Chemical sciences in the 20th century: bridging boundaries. Wiley, Weinheim, pp 14–44
157. Kim M, Press M (2003) Affinity, that elusive dream. Transformations. MIT Press, Cambridge
158. Restrepo G, Villaveces JL (2012) Mathematical thinking in chemistry. Hyle 18:3–22
159. Brodie BC (1866) XXVI. The calculus of chemical operations being a method for the investigation, by means of symbols, of the laws of the distribution of weight in chemical change. Part I. On the construction of chemical symbols. Philos Trans R Soc Lond A 156:781–859
160. Brodie BC (1877) II. The calculus of chemical operations; being a method for the investigation by means of symbols, of the laws of the distribution of weight in chemical change. Part II. on the analysis of chemical events. Philos Trans R Soc Lond 167:35–116
161. Brock WH (2002) Sir Benjamin Collins Brodie (1817–1880). Hyle 8:49–54
162. Bernal A, Llanos E, Leal W, Restrepo G (2015) Similarity in chemical reaction networks: categories, concepts and closure, chap 2. Bentham-Elsevier, Centro Rio de Janeiro, pp 24–54
163. Tabchouri S (2019) A machine learning approach to molecular structure recognition in chemical literature. Master's Thesis, Massachusetts Institute of Technology
164. Staker J, Marshall K, Abel R, McQuaw CM (2019) Molecular structure extraction from documents using deep learning. J Chem Inf Model 59:1017–1029
165. Lavoisier AL (1789) Traité élémentaire de chimie. Accessed 17 Feb 2021
166. Berzelius JJ (1814) Essay on the cause of chemical proportions, and on some circumstances relating to them; together with a short and easy method of expressing them. Ann Philos 3:51–62, 93–106, 244–257, 353–364
167. Woodward RB (1973) The total synthesis of vitamin B12. Pure Appl Chem 33:145–178
168. Trinajstić N (1983) Chemical graph theory. CRC-Press, Boca Raton
169. Cope JD, Sheridan PE, Galloway CJ, Awoyemi RF, Stokes SL, Emerson JP (2020) Synthesis and characterization of a tetradentate, N-heterocyclic carbene copper(II) complex and its use as a Chan–Evans–Lam coupling catalyst. Organometallics 39:4457–4464
170. Nakamuro T, Sakakibara M, Nada H, Harano K, Nakamura E (2021) Capturing the moment of emergence of crystal nucleus from disorder. J Am Chem Soc 143:1763–1767
171. Engel T, Gasteiger J (2018) Chemoinformatics: basic concepts and methods. Wiley, New York

172. Huisgen R, Gilde HG (1979) Rudolf Criegee (1902–1975). J Chem Educ 56:369
173. Coller H (1990) Chemical information 2. Springer, Berlin
174. Sanderson K (2008) Data on display. Nature 455(7211):273
175. Su S, Yang Y, Gan H, Zheng S, Gu F, Zhao C, Xu J (2020) Predicting the feasibility of copper(I)-catalyzed alkyne–azide cycloaddition reactions using a recurrent neural network with a self-attention mechanism. J Chem Inf Model 60:1165–1174
176. Raccuglia P, Elbert KC, Adler PDF, Falk C, Wenny MB, Mollo A, Zeller M, Friedler SA, Schrier J, Norquist AJ (2016) Machine-learning-assisted materials discovery using failed experiments. Nature 533:73–76
177. Nowak MA, Krakauer DC (1999) The evolution of language. Proc Natl Acad Sci 96:8028–8033
178. Lieberman E, Michel JB, Jackson J, Tang T, Nowak MA (2007) Quantifying the evolutionary dynamics of language. Nature 449:713–716
179. Schummer J (2004) Substances versus reactions. Hyle 10:3–4
180. Brock WH (1967) The atomic debates, Brodie and the rejection of the atomic theory. Leicester University Press, Leicester
181. Earley J (1981) Self-organization and agency: in chemistry and in process philosophy. Process Stud 11:242–258
182. Earley JE (1993) The nature of chemical existence. State University of New York Press, New York, pp 272–284
183. Earley J (1998) Modes of chemical becoming. Hyle 4:105–115
184. Earley Sr JE (2003) Varieties of properties. Ann N Y Acad Sci 988:80–89
185. Brakel JV (1997) Chemistry as the science of the transformation of substances. Synthese 111:253–282
186. Stein RL (2004) Towards a process philosophy of chemistry. Hyle 10:5–22
187. Jost J (2015) Mathematical concepts. Springer, Cham
188. Hennig W, Davis D, Zangerl R (1999) Phylogenetic systematics. University of Illinois Press, Champaign
189. Breidbach O, Jost J (2004) Working in a multitude of trends – species balancing populations. J Zool Syst Evol Res 42:202–207
190. Mayr E (2014) Animal species and evolution. Harvard University Press, Cambridge
191. Mayr E (2008) The biological meaning of species. Biol J Linn Soc 1:311–320
192. Scherrer K, Jost J (2007) Gene and genon concept: coding versus regulation. Theory Biosci 126:65–113
193. Nelson D, Lehninger A, Cox M (2008) Lehninger principles of biochemistry. Lehninger principles of biochemistry. W. H. Freeman, New York
194. Reinhardt C (2017) “This other method”: the dynamics of NMR in biochemistry and molecular biology. Hist Stud Nat Sci 47:389–422
195. Chhetri BK, Lavoie S, Sweeney-Jones AM, Kubanek J (2018) Recent trends in the structural revision of natural products. Nat Prod Rep 35:514–531
196. Pye CR, Bertin MJ, Lokey RS, Gerwick WH, Linington RG (2017) Retrospective analysis of natural products provides insights for future discovery trends. Proc Natl Acad Sci 114:5601–5606
197. Yang Z, Jamieson CS, Xue XS, Garcia-Borràs M, Benton T, Dong X, Liu F, Houk KN (2019) Mechanisms and dynamics of reactions involving entropic intermediates. Trends Chem 1:22–34
198. Bruns CJ, Stoddart JF (2016) An introduction to the mechanical bond, chap 1. John Wiley & Sons, Hoboken, pp 1–54
199. Mahjour B, Shen Y, Liu W, Cernak T (2020) A map of the amine-carboxylic acid coupling system. Nature 580:71–75

200. Keserü GM, Soos T, Kappe CO (2014) Anthropogenic reaction parameters - the missing link between chemical intuition and the available chemical space. Chem Soc Rev 43:5387–5399
201. Zhang W, Oganov AR, Goncharov AF, Zhu Q, Boulfelfel SE, Lyakhov AO, Stavrou E, Somayazulu M, Prakapenka VB, Konôpková Z (2013) Unexpected stable stoichiometries of sodium chlorides. Science 342:1502–1505
202. Restrepo G, Harré R (2015) Mereology of quantitative structure-activity relationships models. Hyle 21:19–38
203. Dong X, Oganov AR, Goncharov AF, Stavrou E, Lobanov S, Saleh G, Qian GR, Zhu Q, Gatti C, Deringer VL, Dronskowski R, Zhou XF, Prakapenka VB, Konôpková Z, Popov IA, Boldyrev AI, Wang HT (2017) A stable compound of helium and sodium at high pressure. Nat Chem 9:440–445
204. Morris P (2015) The matter factory. University of Chicago Press, Chicago
205. Stoecklein D, Di Carlo D (2019) Nonlinear microfluidics. Anal Chem 91:296–314
206. Whitesides GM (2006) The origins and the future of microfluidics. Nature 442:368–373
207. Samiei E, Tabrizian M, Hoorfar M (2016) A review of digital microfluidics as portable platforms for lab-on a-chip applications. Lab Chip 16:2376–2396
208. Shestopalov I, Tice JD, Ismagilov RF (2004) Multi-step synthesis of nanoparticles performed on millisecond time scale in a microfluidic droplet-based system. Lab Chip 4:316–321
209. Lindroth E, Calegari F, Young L, Harmand M, Dudovich N, Berrah N, Smirnova O (2019) Challenges and opportunities in attosecond and xfel science. Nat Rev Phys 1:107–111
210. Maiuri M, Garavelli M, Cerullo G (2020) Ultrafast spectroscopy: state of the art and open challenges. J Am Chem Soc 142:3–15
211. Rocke AJ (1993) The quiet revolution. University of California Press, Berkeley
212. Poirier J, Balinski R (1998) Lavoisier: chemist, biologist, economist. Chemical sciences in society. University of Pennsylvania Press, Philadelphia. Incorporated
213. Lavoisier A (2011) Elements of chemistry. Dover Publications, Mineola
214. Thornton BF, Burdette SC (2013) The ends of elements. Nat Chem 5:350–352
215. Smits OR, Mewes JM, Jerabek P, Schwerdtfeger P (2020) Oganesson: a noble gas element that is neither noble nor a gas. Angew Chem Int Ed 59:23636–23640
216. Leal W, Llanos EJ, Stadler PF, Jost J, Restrepo G (2019) The chemical space from which the periodic system arose. ChemRxiv
217. Kekulé A (1865) Sur la constitution des substances aromatiques. Bull Soc Chim Paris 3:98–110
218. Kekulé A (1866) Untersuchungen über aromatische verbindungen ueber die constitution der aromatischen verbindungen. i. ueber die constitution der aromatischen verbindungen. Justus Liebigs Ann Chem 137:129–196
219. Rocke AJ (2015) It began with a daydream: the 150th anniversary of the kekulé benzene structure. Angew Chem Int Ed 54:46–50
220. Kekulé A (1872) Ueber einige condensationsproducte des aldehyds. Justus Liebigs Ann Chem 162:77–124
221. Cosgrove DA, Green KM, Leach AG, Poirrette A, Winter J (2012) A system for encoding and searching markush structures. J Chem Inf Model 52:1936–1947
222. Krchňák V, Holladay MW (2002) Solid phase heterocyclic chemistry. Chem Rev 102:61–92
223. Palomo JM (2014) Solid-phase peptide synthesis: an overview focused on the preparation of biologically relevant peptides. RSC Adv 4:32658–32672
224. Simmons ES (2003) Markush structure searching over the years. World Patent Inf 25:195–202
225. Kerber A, Laue R, Meringer M, Rücker C, Schymanski E (2013) Mathematical chemistry and chemoinformatics: structure generation, elucidation and quantitative structure-property relationships. De Gruyter, Berlin

226. Cankařová N, Schütznerová E, Krchňák V (2019) Traceless solid-phase organic synthesis. Chem Rev 119:12089–12207
227. Latour B (1987) Science in action: how to follow scientists and engineers through society. Harvard University Press, Cambridge
228. Richardson JS (2000) Early ribbon drawings of proteins. Nat Struct Biol 7:624–625
229. Markush structure (2021) https://en.wikipedia.org/wiki/Markush_structure. Accessed 18 Feb 2021
230. Schummer J, Spector TI (2007) The visual image of chemistry: perspectives from the history of art and science. Hyle 13:3–41
231. Centeno SA, Mahon D, Carò F, Pullins D (2021) Discovering the evolution of Jacques-Louis David's portrait of Antoine-Laurent and Marie-Anne Pierrette Paulze Lavoisier. Heritage Sci 9:84
232. Morrison A (1937) Man in a chemical world: the service of chemical industry. Scribner's Sons, New York
233. Ede A (2006) Abraham Cressy Morrison in the Agora: bringing chemistry to the public. Hyle 12:193–214
234. Hepler-Smith E (2016) Nominally rational: systematic nomenclature and the structure of organic chemistry, 1889–1940. Ph.D. Thesis, Princeton University
235. Ball P (2021) Science is political, and we must deal with it. J Phys Chem Lett 12:6336–6340. PMID: 34261323
236. Meija J (2021) Politics at the periodic table. Nat Chem 13:814–816
237. Fontani M, Costa M, Orna M (2015) The lost elements: the periodic table's shadow side. Oxford University Press, Oxford
238. Robinson AE (2021) Attempting neutrality: disciplinary and national politics in a cold war scientific controversy. Centaurus 63:84–102
239. Principe LM (ed) (2007) New narratives in eighteenth-century chemistry. Springer, Berlin
240. Lipkus AH, Yuan Q, Lucas KA, Funk SA, Bartelt WF, Schenck RJ, Trippe AJ (2008) Structural diversity of organic chemistry. A scaffold analysis of the CAS registry. J Org Chem 73:4443–4451
241. Goodnow RA, Dumelin CE, Keefe AD (2017) DNA-encoded chemistry: enabling the deeper sampling of chemical space. Nat Rev Drug Discov 16:131–147
242. Lipinski CA, Litterman NK, Southan C, Williams AJ, Clark AM, Ekins S (2015) Parallel worlds of public and commercial bioactive chemistry data. J Med Chem 58:2068–2076
243. Burnet J (1928) Greek philosophy: thales to Plato. No. pt. 1 in schools of philosophy. Macmillan, London
244. Sarton G (1952) A history of science: ancient science through the golden age of Greece. A history of science. Harvard University Press, Cambridge
245. Sarton G (1959) A history of science: hellenistic science and culture in the last three centuries B.C. A history of science. Harvard University Press, Cambridge
246. Keen E (2007) Journey of a book: bartholomew the Englishman and the properties of things. ANU Press, Canberra
247. Franklin-Brown M (2012) Reading the world: encyclopedic writing in the scholastic age. University of Chicago Press, Chicago
248. Hounshell D, Smith J, Smith J, Galambos L, Gallmam R (1988) Science and corporate strategy: Du Pont R and D, 1902–1980. Studies in economic history and policy: USA in the twentieth century. Cambridge University Press, Cambridge
249. Kurtz SM, Muratoglu OK, Evans M, Edidin AA (1999) Advances in the processing, sterilization, and crosslinking of ultra-high molecular weight polyethylene for total joint arthroplasty. Biomaterials 20:1659–1688

250. Russell MH, Saladini JL, Lichtner F (2002) Sulfonylurea herbicides. Pestic Outlook 13:166–173
251. Bierlein JD, Vanherzeele H (1989) Potassium titanyl phosphate: properties and new applications. J Opt Soc Am B 6:622–633
252. Hoffman R (1984) In memorium: Earl Muetterties. Inorg Chem 23:1789–1789
253. Okazoe T (2009) Overview on the history of organofluorine chemistry from the viewpoint of material industry. Proc Jpn Acad B 85:276–289
254. Webster OW, Hertler WR, Sogah DY, Farnham WB, RajanBabu TV (1983) Group-transfer polymerization. 1. A new concept for addition polymerization with organosilicon initiators. J Am Chem Soc 105:5706–5708
255. Webster OW (2002) Cyanocarbons: a classic example of discovery-driven research. J Polym Sci A Polym Chem 40:210–221
256. Hallonsten O (2016) Big science transformed: science, politics and organization in Europe and the United States. Springer International Publishing, New York
257. Cramer K (2020) A political history of big science: the other Europe. Palgrave studies in the history of science and technology. Springer International Publishing, New York
258. Eagle CT, Sloan J (1998) Marie Anne Paulze Lavoisier: the mother of modern chemistry. Chem Educ 3:1–18
259. Hidden details in iconic portrait of Lavoisiers reveal fears of coming revolution (2021) https://www.chemistryworld.com/opinion/hidden-details-in-iconic-portrait-of-lavoisiers-reveal-fears-of-coming-revolution/4014356.article?utm_source=Nature+Briefing&utm_campaign=f68bb0d42e-briefing-dy-20210913&utm_medium=email&utm_term=0_c9dfd39373-f68bb0d42e-43629497. Accessed 13 Oct 2021
260. Antoine Laurent Lavoisier (1743–1794) and Marie Anne Lavoisier (Marie Anne Pierrette Paulze, 1758–1836) (2021) https://www.metmuseum.org/art/collection/search/436106. Accessed 28 Dec 2021
261. McKie D (1952) Antoine Lavoisier: scientist, economist, social reformer. Life of science library. H. Schuman, New York, NY
262. Simmel G (1908) Soziologie: Untersuchungen über die Formen der Vergesellschaftung. Duncker & Humblot, Leipzig. Reprinted as Suhrkamp Taschenbuch der Wissenschaft, vol 811, 3rd edn., 1999
263. Knorr Cetina K (2002) Wissenskulturen. Suhrkamp, Frankfurt/M
264. Jost J (2014) Mathematical methods in biology and neurobiology. Springer, Cham
265. Painter D, Daniels B, Laubichler M (2019) Innovations are disproportionately likely in the periphery of a scientific network. Theory Biosci 140(4):391–399
266. Scientific publishers expedite name changes for authors (2021). https://www.nature.com/articles/d41586-021-02014-7. Accessed 13 Oct 2021
267. Aftalion F (2001) A history of the international chemical industry. Chemical Heritage Press, Philadelphia
268. Levine I (1983) Physical chemistry. McGraw-Hill, New York
269. Urey HC (1933) Editorial. J Chem Phys 1:1–2
270. Lemonick S (2019) More and more biochemists and life scientists are winning the Nobel Prize in chemistry. Chemical & Engineering News. https://cen.acs.org/people/nobel-prize/biochemists-life-scientists-winning-Nobel/97/web/2019/12. Publication date: December 6, 2019
271. Seeman JI, Restrepo G (2020) The mutation of the "nobel prize in chemistry" into the "nobel prize in chemistry or life sciences": several decades of transparent and opaque evidence of change within the nobel prize program. Angew Chem Int Ed 59:2942–2961

272. List B, Lerner RA, Barbas CF (2000) Proline-catalyzed direct asymmetric aldol reactions. J Am Chem Soc 122:2395–2396
273. Chang H (2009) We have never been whiggish (about phlogiston)1. Centaurus 51:239–264
274. Chang H (2010) The hidden history of phlogiston. how philosophical failure can generate historiographical refinement. Hyle 16:47–79
275. Kuhn TS (1977) The essential tension. The University of Chicago Press, Chicago
276. Partington J (1989) A short history of chemistry. Dover books on chemistry series. Dover Publications, Mineola
277. Fuson RC, Bull BA (1934) The haloform reaction. Chem Rev 15:275–309
278. Jackson CM (2015) The "wonderful properties of glass": Liebig's Kaliapparat and the practice of chemistry in glass. Isis 106:43–69
279. Touchstone JC (1993) History of chromatography. J Liq Chromatogr 16:1647–1665
280. Weil H, Williams TI (1950) History of chromatography. Nature 166:1000–1001
281. Rocke AJ (1978) Atoms and equivalents: the early development of the chemical atomic theory. Hist Stud Phys Sci 9:225–263
282. Council NR (1995) Mathematical challenges from theoretical/computational chemistry. The National Academies Press, Washington, DC
283. Seeman JI, Restrepo G (2020) The mutation of the "Nobel Prize in chemistry" into the "Nobel Prize in chemistry or life sciences": several decades of transparent and opaque evidence of change within the Nobel Prize program. Angew Chem Int Ed 59:2942–2961
284. Friedrich C (ed.) (2022) Friedrich Wilhelm Sertürners Arbeiten zur Entdeckung des Morphins. Klassische Texte der Wissenschaft. Springer, Berlin
285. Gordin MD (2005) Beilstein unbound: the pedagogical unraveling of a man and his handbuch. In: Kaiser D (ed) Pedagogy and the practice of science: historical and contemporary perspectives. The MIT Press, Cambridge, MA, pp 11–39
286. Internet Archive - Chemisches Zentralblatt (2020) https://archive.org/search.php?query=Chemisches%20Zentralblatt. Accessed 6 Apr 2020
287. Wiki-Chemisches Zentralblatt (2020) https://de.wikipedia.org/wiki/Chemisches_Zentralblatt. Accessed 6 Apr 2020
288. Analyzing success rates of supposedly "easy" reactions (2020) https://www.nextmovesoftware.com/products/Filbert4_Boston15.pdf. Accessed 6 Apr 2020
289. Bird CL, Willoughby C, Frey JG (2013) Laboratory notebooks in the digital era: the role of elns in record keeping for chemistry and other sciences. Chem Soc Rev 42:8157–8175
290. Bird CL, Frey JG (2013) Chemical information matters: an e-research perspective on information and data sharing in the chemical sciences. Chem Soc Rev 42:6754–6776
291. De Roure D, Goble C, Aleksejevs S, Bechhofer S, Bhagat J, Cruickshank D, Fisher P, Hull D, Michaelides D, Newman D, Procter R, Lin Y, Poschen M (2010) Towards open science The MyExperiment approach. Concurr Comput Pract Exp 22:2335–2353
292. Bechhofer S, Buchan I, Roure] DD, Missier P, Ainsworth J, Bhagat J, Couch P, Cruickshank D, Delderfield M, Dunlop I, Gamble M, Michaelides D, Owen S, Newman D, Sufi S, Goble C (2013) Why linked data is not enough for scientists. Futur Gener Comput Syst 29:599 – 611. Special section Recent advances in e-Science
293. Herres-Pawlis S, Liermann JC, Koepler O (2020) Research data in chemistry – results of the first NFDI4Chem community survey. Z Anorg Allg Chem 646:1748–1757
294. Koepler O, Liermann J, Schön F, Herres-Pawlis S (2020) Forschungsdatenmanagement - Zeit für den Abschied vom analogen Laborbuch. Nachr Chem 68:20–23
295. Jung N, Neumann S, Koepler O, Bach F, Popp C, Herres-Pawlis S, Liermann J, Razum M, Steinbeck C (2021) NFDI4Chem – Infrastruktur für den digitalen Wandel in der Chemischen Forschung. Bunsen-Magazin 23:100–103

296. Herres-Pawlis S, Koepler O, Steinbeck C (2019) Nfdi4chem: shaping a digital and cultural change in chemistry. Angew Chem Int Ed 58:10766–10768
297. Grinberg N, Rodriguez S (eds) (2019) Ewing's Analytical Instrumentation Handbook. CRC Press, Boca Raton
298. Haight R, Ross FM, Hannon JB (2011) Handbook of instrumentation and techniques for semiconductor nanostructure characterization. World Scientific Publishing Company, Singapore
299. Tsiklis D (2012) Handbook of techniques in high-pressure research and engineering. Springer US, New York
300. Web of science (2020) https://www.webofknowledge.com/. Accessed 6 Apr 2020
301. Dimensions(2020) https://www.dimensions.ai/. Accessed 6 Apr 2020
302. Hook DW, Porter SJ, Herzog C (2018) Dimensions: building context for search and evaluation. Front Res Metrics Anal 3:23
303. CNN-China (2021) Why 1.2 billion people share the same 100 surnames in china. https://edition.cnn.com/2021/01/16/china/chinese-names-few-intl-hnk-dst/index.html. Accessed 25 Oct 2021
304. Roll U, Correia RA, Berger-Tal O (2018) Using machine learning to disentangle homonyms in large text corpora. Conserv Biol 32:716–724
305. Allen L, Scott J, Brand A, Hlava M, Altman M (2009) Credit where credit is due. Nature 462:825–825
306. International institute of social history (2020) https://iisg.amsterdam/en. Accessed 6 Apr 2020
307. Clio infra (2020) https://clio-infra.eu/. Accessed 6 Apr 2020
308. Williams AJ (2008) Internet-based tools for communication and collaboration in chemistry. Drug Discov Today 13:502–506
309. Hamstra D, Kemsley JN, Murray DH, Randall DW (2011) Integrating webinar and blogging technologies into chemistry seminar. J Chem Educ 88:1085–1089
310. Klein DJ (2013) Mathematical chemistry! is it? and if so, what is it? Hyle 19:35–85
311. Restrepo G (2016) Mathematical chemistry, a new discipline, chap 15. Oxford University Press, Oxford, pp 332–351
312. Newman M, Barabási A, Watts DJ (2006) The structure and dynamics of networks. Princeton University Press, Princeton
313. Mulas R, Kuehn C, Jost J (2020) Coupled dynamics on hypergraphs: master stability of steady states and synchronization. Phys Rev E 101:062313
314. Bianconi G (2018) Multilayer networks: structure and function. Oxford university press, Oxford
315. Jost J, Mulas R (2019) Hypergraph Laplace operators for chemical reaction networks. Adv Math 351:870–896
316. Johnson J (2012) Hypernetworks in the science of complex systems. Imperial College Press, London
317. Joharinad P, Jost J (to appear, 2023) Mathematical principles of topological and geometric data analysis. Mathematics of data
318. Leal W, Restrepo G, Stadler PF, Jost J (2018). Forman-Ricci curvature for hypergraphs. Adv Complex Syst 24(01), 2150003
319. Leal W, Eidi M, Jost J (2019) Curvature-based analysis of directed hypernetworks. In: Cherifi H (ed) Complex networks 2019: the 8th international conference on complex networks and their applications; December 10–12, 2019. Lisbon, Portugal, pp 32–34. Book of abstract: international Conference on Complex Networks & Their Applications, [s.l.]
320. Klamt S, Haus UU, Theis F (2009) Hypergraphs and cellular networks. PLOS Comput Biol 5:1–6

321. Bernal A, Daza E (2011) Metabolic networks: beyond the graph. Curr Comput Aided Drug Des 7:122–132
322. Eidi M, Jost J (2019) Ollivier Ricci curvature of directed hypergraphs. Sci Rep 10:12466
323. Leal W, Restrepo G (2019) Formal structure of periodic system of elements. Proc R Soc A Math Phys Eng Sci 475:20180581
324. Bernadette BV (2003) Languages in chemistry. In: Mary JN (ed) The Cambridge history of science, vol 5. Cambridge University Press, Cambridge, pp 174–190
325. Jost J (2007) Dynamical networks. Springer London, London, pp 35–62
326. Joslyn CA, Aksoy S, Callahan TJ, Hunter LE, Jefferson B, Praggastis B, Purvine EAH, Tripodi IJ (2020) Hypernetwork science: from multidimensional networks to computational topology. arXiv preprint. https://doi.org/10.48550/ARXIV.2003.11782. https://arxiv.org/abs/2003.11782
327. Eidi M, Farzam A, Leal W, Samal A, Jost J (2020) Edge-based analysis of networks: curvatures of graphs and hypergraphs. Theory Biosci 139:337–348
328. Farzam A, Samal A, Jost J (2020) Degree difference: a simple measure to characterize structural heterogeneity in complex networks. Sci Rep 10:21348
329. Barabási AL, Albert R (1999) Emergence of scaling in random networks. Science 286:509–512
330. Ay N, Olbrich E, Bertschinger N, Jost J (2011) A geometric approach to complexity. Chaos Interdisciplinary J Nonlinear Sci 21:037103
331. Epstein JM, Axtell RL (1996) Growing artificial societies. Brookings Institution Press/MIT Press, Washington, D.C./Cambridge
332. Ay N, Jost J, Lê HV, Schwachhöfer LJ (2017) Information geometry, vol. 64 of Ergebnisse der Mathematik und ihrer Grenzgebiete. 3. Folge / A series of modern surveys in mathematics. Springer, Cham
333. Ay N, Krakauer DC (2007) Geometric robustness theory and biological networks. Theory Biosci 125:93–121
334. Kantz H, Schreiber T (1997) Nonlinear time series analysis. Cambridge University Press, Cambridge
335. Fulcher BD, Little MA, Jones NS (2013) Highly comparative time-series analysis: the empirical structure of time series and their methods. J R Soc Interface 10:20130048
336. Wu Z, Huang NE, Long SR, Peng CK (2007) On the trend, detrending, and variability of nonlinear and nonstationary time series. Proc Natl Acad Sci 104:14889–14894
337. Brillinger D (2001) Time series: general. In: Smelser NJ, Baltes PB (eds) International encyclopedia of the social & behavioral Sciences. Pergamon, Oxford, pp 15724 – 15731
338. Laubichler MD, Maienschein J, Renn J (2019) Computational history of knowledge: challenges and opportunities. Isis 110:502–512
339. Wolpert DH, Price MH, Crabtree SA, Kohler TA, Jost J, Evans J, Stadler PF, Shimao H, Laubichler MD (2021) The past as a stochastic process. arXiv preprint arXiv:211205876
340. PNAS Moves to a Continuous Publication Model for 2021 (2021) https://www.pnas.org/page/updates#nexus-launch. Accessed 1 Dec 2021
341. Aleta A, Moreno Y (2019) Multilayer networks in a nutshell. Ann Rev Condens Matter Phys 10:45–62
342. Danziger MM, Bonamassa I, Boccaletti S, Havlin S (2019) Dynamic interdependence and competition in multilayer networks. Nat Phys 15:178–185
343. Chodrow P, Mellor A (2020) Annotated hypergraphs: models and applications. Appl Netw Sci 5:9
344. Kaneko K (1984) Period-doubling of kink-antikink patterns, quasiperiodicity in antiferro-like structures and spatial intermittency in coupled logistic lattice: towards a prelude of a "field theory of chaos". Prog Theor Phys 72:480–486

345. Pikovsky A, Rosenblum M, Kurths J (2001) Synchronization. Cambridge University Press, Cambridge
346. Atay FM, Jost J (2004) On the emergence of complex systems on the basis of the coordination of complex behaviors of their elements. Complexity 10:17–22
347. Lu W, Atay FM, Jost J (2007) Synchronization of discrete-time dynamical networks with time-varying couplings. SIAM J Math Anal 39:1231–1259
348. Vespignani A (2012) Modelling dynamical processes in complex socio-technical systems. Nat Phys 8:32–39
349. Feng M, Cai SM, Tang M, Lai YC (2019) Equivalence and its invalidation between non-Markovian and Markovian spreading dynamics on complex networks. Nat Commun 10:3748
350. Rosati DP, Woolhouse MH, Bolker BM, Earn DJD (2021) Modelling song popularity as a contagious process. Proc R Soc A Math Phys Eng Sci 477:20210457
351. Schummer J (1999). Coping with the growth of chemical knowledge: challenges for chemistry documentation, education, and working chemists. Educación Química 10(2):92–101
352. Levantovsky R, Vabret N (2020) Hydroxychloroquine: small effects in mild disease. Nat Rev Immunol 20(6):350
353. Wang M, Cao R, Zhang L, Yang X, Liu J, Xu M, Shi Z, Hu Z, Zhong W, Xiao G (2020) Remdesivir and chloroquine effectively inhibit the recently emerged novel coronavirus (2019-nCoV) in vitro. Cell Res 30:269–271
354. Bonabeau E (2002) Agent-based modeling: methods and techniques for simulating human systems. Proc Natl Acad Sci 99:7280–7287
355. Hommes C (2013) Behavioral rationality and heterogeneous expectations in complex economic systems. Cambridge University Press, Cambridge
356. Kohler TA, Gumerman GG (2000) Dynamics in human and primate societies: agent-based modeling of social and spatial processes. Oxford University Press, Oxford
357. Wurzer G, Kowarik K, Reschreiter H (2014) Agent-based modeling and simulation in archaeology. Advances in Geographic Information Science. Springer International Publishing, Berlin
358. Turchin P, Nefedov SA (2009) Secular cycles. Princeton University Press, Princeton
359. Fialkowski M, Bishop KJM, Chubukov VA, Campbell CJ, Grzybowski BA (2005) Architecture and evolution of organic chemistry. Angew Chem Int Ed 44:7263–7269
360. Bensaude-Vincent B (1990) A view of the chemical revolution through contemporary textbooks: Lavoisier, Fourcroy and Chaptal. Br J Hist Sci 23:435–460
361. Sukopp T (2018) Discoveries of oxygen and the "chemical revolution" in the context of European scientific networks. Springer Fachmedien Wiesbaden, Wiesbaden, pp 15–47
362. Chang H (2015) The chemical revolution revisited. Stud Hist Philos Sci A 49:91 – 98
363. Altschul JH, Kintigh KW, Klein TH, Doelle WH, Hays-Gilpin KA, Herr SA, Kohler TA, Mills BJ, Montgomery LM, Nelson MC, Ortman SG, Parker JN, Peeples MA, Sabloff JA (2017) Opinion: fostering synthesis in archaeology to advance science and benefit society. Proc Natl Acad Sci 114:10999–11002
364. Nunn N (2020) The historical roots of economic development. Science 367
365. Teschendorff AE (2019) Avoiding common pitfalls in machine learning omic data science. Nat Mater 18:422–427
366. Eberlen J, Scholz G, Gagliolo M (2017) Simulate this! An introduction to agent-based models and their power to improve your research practice. Int Rev Soc Psychol 30:149–160
367. Matín-Vide C (2003) Formal grammars and languages. In: Mitkov R (ed) The Oxford handbook of computational linguistics. Oxford University Press, Oxford, pp 157–177

368. Andersen JL, Flamm C, Merkle D, Stadler PF (2016) A software package for chemically inspired graph transformation. In: Echahed R, Minas M (eds) Graph transformation. Springer International Publishing, Cham, pp 73–88
369. Andersen JL, Flamm C, Merkle D, Stadler PF (2017). Chemical transformation motifs - modelling pathways as integer hyperflows. IEEE/ACM Trans Comput Biol Bioinform 16(2):510-523
370. Quintero NY, Restrepo G (2017) Formal concept analysis applications in chemistry: from radionuclides and molecular structure to toxicity and diagnosis. Springer International Publishing, Cham, pp 207–217
371. Restrepo G (2020) A formal approach to the conceptual development of chemical element. Oxford University Press, New York, pp 225–240
372. Ganter B, Wille R (1999) Formal concept analysis. Springer, Berlin
373. Gärdenfors P, Jost J, Warglien M (2018) From actions to effects: three constraints on event mappings. Front Psychol 9:1391
374. Bengfort B, Bilbro R, Ojeda T (2018) Applied text analysis with Python: enabling language-aware data products with machine learning. O'Reilly Media, Sebastopol
375. Føllesdal D (1997) Semantics and semiotics. Springer Netherlands, Dordrecht, pp 449–457
376. Nöth W (2000) Handbuch der Semiotik. J.B.Metzler, Stuttgart/Weimar
377. Suppe F (1998) The structure of a scientific paper. Philos Sci 65:381–405
378. Cachola I, Lo K, Cohan A, Weld DS (2020) TLDR: extreme summarization of scientific documents. arXiv preprint. https://doi.org/10.48550/ARXIV.2004.15011. https://arxiv.org/abs/2004.15011
379. Shackell C, Sitbon L (2019) Computational opposition analysis using word embeddings: a method for strategising resonant informal argument. Argument Comput 10(3):301–317
380. Assaf D, Cohen Y, Danesi M, Neuman Y (2015) Opposition theory and computational semiotics. Sign Syst Stud 43:159–172
381. Chartier JF, Pulizzotto D, Chartrand L, Meunier JG (2019) A data-driven computational semiotics: the semantic vector space of Magritte's artworks. Semiotica 2019:19–69
382. De Saussure F (1989) Cours de linguistique générale, vol 1. Otto Harrassowitz Verlag, Wiesbaden
383. Widyassari AP, Rustad S, Shidik GF, Noersasongko E, Syukur A, Affandy A, Setiadi DRIM (2022) Review of automatic text summarization techniques & methods. J King Saud Univ Comput Inf Sci 34(4):1029–1046
384. Cederman LE, Weidmann NB (2017) Predicting armed conflict: time to adjust our expectations? Science 355:474–476
385. Beck N, King G, Zeng L (2000) Improving quantitative studies of international conflict: a conjecture. Am Polit Sci Rev 94:21–36
386. Molina M, Garip F (2019) Machine learning for sociology. Ann Rev Sociol 45:27–45
387. Imai K, Ratkovic M (2013) Estimating treatment effect heterogeneity in randomized program evaluation. Ann Appl Stat 7:443–470
388. DiMaggio P, Nag M, Blei D (2013) Exploiting affinities between topic modeling and the sociological perspective on culture: application to newspaper coverage of U.S. government arts funding. Poetics 41:570–606. Topic Models and the Cultural Sciences
389. Gómez-Bombarelli R, Aguilera-Iparraguirre J, Hirzel TD, Duvenaud D, Maclaurin D, Blood-Forsythe MA, Chae HS, Einzinger M, Ha DG, Wu T, Markopoulos G, Jeon S, Kang H, Miyazaki H, Numata M, Kim S, Huang W, Hong SI, Baldo M, Adams RP, Aspuru-Guzik A (2016) Design of efficient molecular organic light-emitting diodes by a high-throughput virtual screening and experimental approach. Nat Mater 15:1120–1127

390. Segler MHS, Waller MP (2017) Modelling chemical reasoning to predict and invent reactions. Chem A Eur J 23:6118–6128
391. Segler MHS, Preuss M, Waller MP (2018) Planning chemical syntheses with deep neural networks and symbolic AI. Nature 555:604–610
392. Cova TFGG, Pais AACC (2019) Deep learning for deep chemistry: optimizing the prediction of chemical patterns. Front Chem 7:809
393. Mater AC, Coote ML (2019) Deep learning in chemistry. J Chem Inf Model 59:2545–2559
394. Naturebias (2020) Bias detectives: the researchers striving to make algorithms fair. https://www.nature.com/articles/d41586-018-05469-3. Accessed 26 Nov 2020
395. Li J, Liu L, Le TD, Liu J (2020) Accurate data-driven prediction does not mean high reproducibility. Nat Mach Intell 2:13–15
396. Jost J (2017) Object oriented models vs. data analysis - is this the right alternative? In: Lenhard J, Carrier M (eds.) Mathematics as a tool: tracing new roles of mathematics in the sciences. Boston studies in the philosophy and history of science, vol 327. Springer, Berlin, pp 253–286
397. Pearl J (2009) Causality. Cambridge University Press, Cambridge
398. Schölkopf B (2019) Causality for machine learning. arXiv preprint. https://doi.org/10.48550/ARXIV.1911.10500. https://arxiv.org/abs/1911.10500
399. Zhang C, Bengio S, Hardt M, Recht B, Vinyals O (2016) Understanding deep learning requires rethinking generalization. arXiv preprint arXiv:161103530
400. Belkin M, Hsu D, Ma S, Mandal S (2019) Reconciling modern machine-learning practice and the classical bias–variance trade-off. Proc Natl Acad Sci 116:15849–15854
401. Anastas PT (2019) Beyond reductionist thinking in chemistry for sustainability. Trends Chem 1:145–148
402. Santa Fe Institute (2020). https://santafe.edu/engage/learn/schools. Accessed 6 Apr 2020
403. Ashby I, Exter M (2019) Designing for interdisciplinarity in higher education: considerations for instructional designers. TechTrends 63:202–208
404. Schummer J (1997) Scientometric studies on chemistry ii: aims and methods of producing new chemical substances. Scientometrics 39:125–140
405. Kowalczyk S, Shankar K (2018) Data sharing and the future of science. Nat Commun 9:2817
406. Data sharing and how it can benefit your scientific career (2020). https://www.nature.com/articles/d41586-019-01506-x. Accessed 6 Apr 2020
407. O'Boyle NM, Guha R, Willighagen EL, Adams SE, Alvarsson J, Bradley JC, Filippov IV, Hanson RM, Hanwell MD, Hutchison GR, James CA, Jeliazkova N, Lang AS, Langner KM, Lonie DC, Lowe DM, Pansanel J, Pavlov D, Spjuth O, Steinbeck C, Tenderholt AL, Theisen KJ, Murray-Rust P (2011) Open data, open source and open standards in chemistry: the blue obelisk five years on. J Cheminf 3:37
408. Gell-Mann M (1995) The quark and the jaguar: adventures in the simple and the complex. W. H. Freeman, New York

Index

J. Jost, G. Restrepo, *The Evolution of Chemical Knowledge*, Wissenschaft und Philosophie – Science and Philosophy – Sciences et Philosophie,
https://doi.org/10.1007/978-3-031-10094-9

Made in the USA
Monee, IL
13 September 2026

62973716R00077